Engineering Mechanics
for Structures

Engineering Mechanics for Structures

Louis L. Bucciarelli
Massachusetts Institute of Technology

DOVER PUBLICATIONS
Garden City, New York

Bibliographical Note

This Dover edition, first published in 2009, is the first publication in book form of a course text prepared for Solid Mechanics, MIT Civil and Environmental Engineering Course #1.050, for Fall 2004. The Preface has been specially updated for the present edition.

Library of Congress Cataloging-in-Publication Data

Bucciarelli, Louis L.
    Engineering mechanics for structures / Louis L. Bucciarelli.
        p. cm.
    ISBN-13: 978-0-486-46855-6
    ISBN-10: 0-486-46855-0
        1. Structural analysis (Engineering). I. Title.

TA645.B74 2009
624.1'7—dc22

2008037649

Manufactured in the United States of America
4500056416
www.doverpublications.com

Preface

This Dover Edition of *Engineering Mechanics for Structures* is intended for use by students of engineering and architecture. It provides an introduction to the fundamental concepts and principles governing the static, elastic behavior of structural elements and forms, e.g., trusses, beams and frames. It may find use in a course in Solid Mechanics as well as in the more universally required subject labeled Statics and Strength of Materials - the latter normally engaged at the sophomore level by students majoring in Mechanical, Civil, or Aeronautical Engineering. My main motivation for writing on such a common, entry level, engineering subject - one already supported by several excellent and enduring textbooks - e.g., Crandall, Dahl & Lardner, *Solid Mechanics*, and Beer, F. P., Johnston, E. R. Jr., & DeWolf, J. T., *Mechanics of Materials* - was a felt need to bridge the gap between the analysis of structures and the engineering design of the same.

Throughout the text I have tried to break free of the linear, reductive, strictly logical approach to problem solving, showing how the way an engineer constructs a problem and attempts a solution is not unique; a problem can be defined and worked in different ways. This is particularly true in designing. (Note: "Design" is to be understood broadly as encompassing all stages in the development of a product or system from reading and interpretation of specifications, through conceptualization of alternatives, prototyping, detailing, market testing and the like[1]). The design exercises included are 'open-ended'; they admit of, not only more than a single right answer, but can be addressed via alternate methods. The value of this, pedagogically speaking, is that students must appropriate the problem as their own. Generally, once they get over the feelings of insecurity of being cast adrift, free of the game of finding the right equation to plug-into, they take to the challenge of refining specifications, estimating the magnitudes of significant parameters, developing an appropriate model for analysis, and evaluating their results.

It is difficult to convey in a cryptic statement of a design exercise, as found at the end of each chapter, all of the contingencies and constraints that might guide and govern the student's efforts in addressing the task; there are too many different ways in which one might set an open-ended, design task. So it becomes the teacher's responsibility to shape and define the classroom context according to their own interests and understanding of engineering practice (and according to time available). Evaluating the students' efforts is also a chore; to read their work - I have students keep a journal - and to appreciate as well as judge each individual's approach takes time; certainly more time than grading a single-right-answer problem set. But I have found that it is well worth the effort. For students find these exercises (six throughout the semester) the most valuable part of the course yet, at the same time, the most time and energy consuming.

In some ways the text is not complete - e.g., the chapter on failure mechanisms is but cursory - but it is my belief, reinforced over a decade spent developing and teaching this material, that more important than coverage is to instill in students the spirit of the engineering enterprise with all of its uncertainties as well as opportunities for creative making. Readers of this Dover Edition might find useful some computer tools for the analysis of truss and frame structures and a few interactive exercises available online at MIT's Open CourseWare site under the heading Civil Engineering, course 1.050.

Louis L. Bucciarelli
24 May, 2008

---

1. For a fuller exposition of the author's vision of engineering design, see *Designing Engineers,* MIT Press, 1994.

Contents

1

Introduction

1.1 What's it about?

This is a book about the Mechanics of Solids, Statics, the Strength of Materials, and Elasticity Theory. But that doesn't mean a thing unless you have had a course in the Mechanics of Solids, Statics, the Strength of Materials, or Elasticity Theory. I assume you have not; let us try again:

This is a book that builds upon what you were supposed to learn in your basic physics and mathematics courses last year. We will talk about forces – not political, but vector forces – about moments and torques, reactions, displacements, linear springs, and the requirements of static equilibrium of a particle or a rigid body. We will solve sets of linear algebraic equations and talk about when we can not find a unique solution to a set of linear algebraic equations. We will derive a whole raft of new equations that apply to particles, bodies, structures, and mechanisms; these will often contain the spatial derivatives of forces, moments, and displacements. You have seen a good bit of the basic stuff of this course before, but we will not assume you know the way to talk about, or work with, these concepts, principles, and methods so fundamental to our subject. So we will recast the basics in our own language, the language of engineering mechanics.

For the moment, think of this book as a language text; of yourself as a language student beginning the study of Engineering Mechanics, the Mechanics of Solids, the Strength of Materials, and Elasticity Theory. You must learn the language if you aspire to be an engineer. But this is a difficult language to learn, unlike any other foreign language you have learned. It is difficult because, on the surface, it appears to be a language you already know. That is deceptive: You will have to be on guard, careful not to presume the word you have heard before bears the same meaning. Words and phrases you have already encountered now take on a more special and, in most cases, narrower meaning; a *couple of forces* is more than just two forces.

An important part of learning the vocabulary, is the quick sketch. Along with learning to sketch in the engineering mechanics way, you will have to learn the meaning of certain icons; a small circle, for example, becomes a frictionless pin. So too, grammar and syntax will be crucial. Rigorous rules must be learned and obeyed. Some of these rules will at first seem pedantic; they may strike you as not only irrelevant to solving the problem, but wrong-headed or counter intuitive. But don't despair; with use they will become familiar and reliable friends.

When you become able to speak and respond in a foreign language without thinking of every word, you start to see the world around you from a new perspective. What was once a curiosity now is mundane and used everyday, often without thinking. So too, in this course, you will look at a tree and see its limbs as cantilever beams, you will look at a beam and see an internal bending moment, you will look at a bending moment and conjecture a stress distribution. You will also be asked to be creative in the use of this new language, to model, to estimate, to design.

That's the goal: To get you seeing the world from the perspective of an engineer responsible for making sure that the structure does not fail, that the mechanism doesn't make too much noise, that the bridge doesn't sway in the wind, that the latch latches firmly, that the landing gear doesa not collapse upon touchdown, the drive-shaft does not fracture in fatigue... Ultimately, that is what this book is about. Along the way you will learn about stress, strain, the behavior of trusses, beams, of shafts that carry torsion, even columns that may buckle.

1.2   What you will be doing.

The best way to learn a foreign language is from birth; but then it's no longer a foreign language. The next best way to learn a new language is to use it – speak it, read it, listen to it on audio tapes, watch it on television; better yet, go to the land where it is the language in use and use it to buy a loaf of bread, get a hotel room for the night, ask to find the nearest post office, or if you are really proficient, make a telephone call. So too in the Mechanics of Solids, we insist you begin to use the language.

Doing problems and exercises, taking quizzes and the final, is using the language. This book contains mostly exercises explained as well as exercises for you to tackle. There are different kinds of exercises, different kinds for the different contexts of language use.

Sometimes an engineer will be asked a question and a response will be expected in five minutes. You will not have time to go to the library, access a database, check this textbook. You must estimate, conjure up a rational response on the spot. "....How many piano tuners in the city of Chicago?" (Try it!) Some of the exercises that follow will be labeled *estimate*.

Often practicing engineers must ask what they need to know in order to tackle the task they have been assigned. So too we will ask you to step back from a problem and pose a new problem that will help you address the original problem. We will label these exercises *need to know*.

A good bit of engineering work is variation on a theme, changing things around, recasting a story line, and putting it into your own language for productive and profitable use. Doing this requires experimentation, not just with hardware, but with concepts and existing designs. One poses "what if we make this strut out of aluminum... go to a cantilever support... pick up the load in bending?" We will label this kind of exercise *what if?*

Engineering analysis (as well as prototype testing and market studies) is what justifies engineering designs. As an engineer you will be asked to show that your design will actually work. More specifically, in the terms of this subject, you will be asked to show that the requirements of static equilibrium ensure your proposed structure will bear the anticipated loading, that the maximum deflection of simply supported beam at midspan does not exceed the value specified in the contract, that the lowest resonant frequency of the payload is above 100 Hz. The *show that* label indicates a problem where a full analysis is demanded. Most often this kind of problem will admit of a single solution – in contrast to the *estimate, need to know*, or even *what if* exercise. This is the form of the traditional textbook problem set.

Closely related to *show that* exercises, you will be asked to *construct* a response to questions, as in "*construct* an explanation explaining why the beam failed", or "*construct* an expression for the force in member ab". We could have used the ordinary language, "explain" or "derive" in these instances but I want to emphasize the initiative you, as learner, must take in explaining or deriving. Here, too, *construct* better reflects what engineers actually do at work.

Finally, what engineers do most of the time is design, design in the broadest sense of the term; they play out scenarios of things working, construct stories and plans that inform others how to make things that will work according to their plans. These *design exercises* are the most open-ended and unconstrained exercises you will find in this text. We will have more to say about them later.

In working all of these different kinds of exercises, we want you to use the language with others. Your one-on-one confrontation with a problem set or an exam question can be an intense dialogue but it's not full use of the language. Your ability to speak and think in the language of engineering mechanics is best developed through dialogue with your peers, your tutors, and your teacher. We encourage you to learn from your classmates, to collectively learn from each other's mistakes and questions as well as problem-solving abilities.

Exercise 1.1 An Introductory Exercise

Analyze the behavior of the mechanism shown in the figure. That is, determine how the deflection, Δ, of point B varies as the load, P, varies.

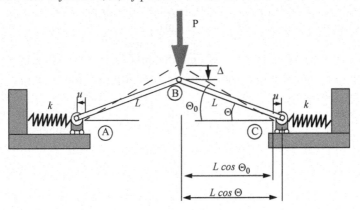

This exercise is intended to introduce the essential concepts and principles of the engineering mechanics of solids. It is meant as an overview; do not be disturbed by the variety of concepts or range of vocabulary. We will try to grasp the essential workings of the device and begin to see the relevance of the concepts and principles of engineering mechanics to an understanding of how it functions and how it might be made to work better.

We will apply *the requirements of static equilibrium.* We will analyze the *displacements* of different points of the structure, e.g., the vertical displacement of point B and the horizontal displacements of points A and C, and make sure these are *compatible.* We will consider the *deformation* of the springs which connect points A and C to ground and posit a relationship between the force each one bears and the *relative displacement* of one end with respect to the other end. We will assume this *force/deformation relationship* is *linear.*

We will learn to read the figure; k is the constant of proportionality in the equation relating the force in the spring to its deformation; the little circles are *frictionless pins,* members AB and BC are *two-force members* — as straight members they carry only *tension and compression.* The grey shading represents *rigid ground.* The arrow represents a vertically *applied load* whose *magnitude* is "P".

Our aim is to determine the *behavior of the structure* as the applied load increases from zero to some or any finite value. In particular, we want to determine how Δ varies as P changes. We will use a spread sheet to make a graph of this relationship — but only after setting up the problem in terms of *non-dimensional expressions* for the applied load, P, the vertical displacement, Δ, and the horizontal displacement, u. We will allow for relatively *large displacements and rotations.* We will investigate the possibility of *snap-through,* a type of *instability,* if P gets too large. In sum, our objective is to determine how the applied load P varies with Δ, or, alternatively, for any prescribed Δ, what P need be applied?

We will discuss how this funny looking linkage of impossible parts (frictionless pins, rollers, rigid grey matter, point loads, ever linear springs) can be a useful model of real-world structures. There is much to be said; all of this italicized language is important.

We start by reasoning thus:

Clearly the vertical displacement, Δ, is related to the horizontal displacements of points A and C; as these points move outward, point B moves downward. We assume our system is *symmetric;* the figure suggests this; if A moves out a distance u, C displaces to the right the same distance. Note: Both u and Δ are measured from the *undeformed* or *unloaded configuration,* P, $\Delta = 0$. (This undeformed configuration is indicated by the dashed lines and the angle Θ_0). As P increases, Δ increases and so too u which causes the springs to shorten. This engenders a compressive force in the springs and in the members AB and BC, albeit of a different magnitude, which in turn, ensures static equilibrium of the system and every point within it including the node B where the load P is applied. But enough talk; enough story telling. We formulate some equations and try to solve them.

Static Equilibrium of Node B.

The figure at the right shows an *isolation* of node B. It is a *free body diagram*; i.e., the node has been cut free of all that surrounds it and the influence of those surroundings have been represented by forces. The node is compressed by the force, F, carried by the members AB and BC. We defer a proof that the force must act along the member to a later date.

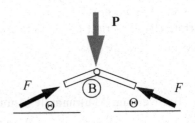

Equilibrium requires that *the resultant force* on the node vanish; symmetry, with respect to a vertical plane containing P and perpendicular to the page, assures this requirement is satisfied in the horizontal direction; equilibrium in the vertical direction gives:

$$2F \cdot sin\theta - P = 0 \qquad \text{or} \qquad F = P/(2\,sin\theta)$$

Static Equilibrium of Node C.

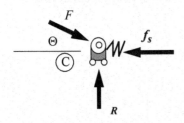

The figure at the right shows an isolation of node C. Note how I have drawn F in this isolation acting opposite to the direction of F shown in the isolation of node B. This is because member BC is in compression. The member is compressing node C as it is compressing node B. R is the *reaction force* on the node *due to the ground*. It is vertical since the rollers signify that there is no *resistance to motion* in the horizontal direction; there is no *friction*. f_s is the compressive force in the spring, again, pressing on the node C.

Equilibrium requires that the *resultant* of the three forces vanish. Requiring that the sum of the horizontal components and that the sum of the vertical components vanish independently will ensure that the vector sum, which is the resultant, will vanish. This yields two scalar equations:

$$F \cdot cos\theta - f_s = 0 \qquad \text{and} \qquad R - F \cdot sin\theta = 0$$

The first ensures equilibrium in the horizontal direction, the second, in the vertical direction.

Force/Deformation of the spring:

For our *linear spring*, we can write:

$$f_s = k \cdot u$$

The force is proportional to deformation.

Compatibility of Deformation:

Δ and Θ are not *independent*; you can not choose one arbitrarily, then the other arbitrarily. The first figure indicates that, if members AB and BC remain *continuous* and rigid[1], we have

$$u = L(cos\theta - cos\theta_0)$$

But we want Δ, ultimately. Another *compatibility relationship* is seen to be
$$\Delta = L(sin\theta_0 - sin\theta)$$

Solution: The relation between *P* and Δ.

From equilibrium of node B we have an equation for *F*, the compressive force in members AB and BC in terms of *P* and θ; Using our result obtained from equilibrium of node C, we can express the force in the spring, f_s in terms of *P* and θ.
$$f_s = (P/2)(cos\theta/sin\theta)$$
From the force/deformation relationship for the spring we express, *u*, the displacement of the node C (and A) in terms of f_s, and, using the immediately above, in terms of *P* and θ.
$$u = f_s/k = (P/2k)(cos\theta/sin\theta)$$
The first compatibility relationship then allows us to write

$$(P/2kL) = sin\theta \cdot (1 - cos\theta_0/cos\theta)$$

while the second may be written in *non-dimensional form* as

$$\Delta/L = sin\theta_0 - sin\theta$$

We could at this point try to eliminate theta; but this is unnecessary. This parametric form of the relationship between *P* and Δ will suffice. Theta serves as an intermediary - a parameter whose value we can choose - guided by our sketch of the geometry of our structure. For each value of theta, the above two equations then fix the value of the vertical displacement and the applied load. A spread sheet is an appropriate tool for carrying out a sequence of such calculations and for constructing a graph of the way *P* varies with Δ, which is our objective. This is left as an exercise for the reader.

---

1. They neither lengthen nor contract when loaded.

1.3 Resources you may use.

A textbook is only one resource available to you in learning a new language. The exercises are another, pehaps the most important other resource you have available. Still others are the interactive short simulations – computer representations of specific problems or phenomena – made available to you over the web. You will find there as well more sophisticated and generally applicable tools which will enable you to model truss and frame structures - structures which have many members. Another more standard and commonly available tool is the spreadsheet. You will find all these modeling tools to be essential and powerful aids when confronted with an open-ended design exercise where the emphasis is on *what if* and *show that*.

Another resource to you is your peers. We expect you to learn from your classmates, to collaborate with them in figuring out how to set up a problem, how to use a spreadsheet, where on the web to find a useful reference. Often you will be asked to work in groups of two or three, in class - especially when a design exercise is on the table - to help formulate a specification and flesh out the context of the exercise. Yet your work is to be your own.

1.4   Problems

1.1 Without evaluating specific numbers, sketch what you think a plot of the load P - in nondimensional form - $(P/2kL)$ versus the displacement, (Δ/L) will look like for the Introductory Exercise above. Consider θ to vary over the interval $\Theta_0 \geq \Theta \geq -\Theta_0$.

Now compute, using a spreadsheet, values for load and displacement and plot over the same range of the parameter Θ.

2

Static Equilibrium Force and Moment

2.1   Concept of Force

Equilibrium of a Particle

You are standing in an elevator, ascending at a constant velocity, what is the resultant force acting on you as a particle?

The correct response is zero: **For a particle at rest, or moving with constant velocity relative to an inertial frame, the resultant force acting on the *isolated particle* must be zero, must vanish.** We usually attribute this to the unquestionable authority of Newton.

The essential phrases in the question are *constant velocity, resultant force* and *particle.* Other words like "standing", "elevator", "ascending", and "you" seem less important, even distracting, but they are there for a reason: The world that you as an engineer will analyze, re-design, and systematize is filled with people and elevators, not isolated particles, velocity vectors, or resultant forces — or at least, not at first sight. The latter concepts are abstractions which you must learn to identify in the world around you in order to work effectively as an engineer, e.g., in order to design an elevator. The problems that appear in engineering text books are a kind of middle ground between abstract theory and everyday reality. We want you to learn to read and see through the superficial appearances, these descriptions which mask certain scientific concepts and principles, in order to grasp and appropriate the underlying forms that provide the basis for engineering analysis and design.

The key phrase in Newton's requirement is ***isolated particle***: It is absolutely essential that you learn to abstract out of the problem statement and all of its relevant and irrelevant words and phrases, a vision of a particle as a point free in space. It's best to render this vision, this abstraction "hard" by drawing it on a clean sheet of paper. Here is how it would look.

●

An Isolated Particle:

You, in an elevator.

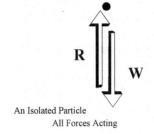

An Isolated Particle
All Forces Acting

This is a non-trivial step, akin to a one month old's apprehension that there are other egos in the world. You are to take the dot drawn as the representation of a thing, all things, that can be thought of as an isolated particle.

Now show all the forces acting on the particle. We have the force due to gravity, $W=Mg$, acting vertically down, toward the center of the earth.... (Who said the elevator was oriented vertically? Who said it was on the surface of the earth? This information is not given; indeed, you could press the point, arguing that the question is not well posed. But is this information essential? We return to this point at the end of this chapter). We have the *reaction force* of the elevator floor acting vertically upward on you, on you as an abstraction, as an isolated particle. This is how our particle looks with all forces acting upon it.

The *resultant force* is the vector sum of all the forces acting on the isolated particle. For *static equilibrium* of the isolated particle, the resultant of the two forces – W acting downward and R acting upward – must be zero.

$$R - W = 0$$

This leads to the not very earth shaking conclusion that the magnitude of the reaction force, acting up, must equal the weight.

$$R = W$$

This apparently trivial result and simplicity of the problem, if indeed it can be called a problem, ought not to be allowed to deceive us: The introduction of the reaction of the floor on you, the passenger in the elevator, is characteristic of the most difficult step in applying the requirement of static equilibrium to an isolated particle. You will find it takes courage, as well as facility with the language of engineering mechanics, to venture forth and construct reaction forces out of thin air. They are there, hidden at the interface of your particle with the rest of the world. Some, like gravity, act at a distance, across all boundaries you may draw.

Exercise 2.1

Estimate the lift force acting on the wings of a Boeing 747 traveling from New York to Los Angeles during rush hour.

We can use the same isolation, or **free-body diagram,** of the figures above where now the point represents the Boeing 747, rather than you in an elevator, and the reaction force represents the lift force acting on the airplane, rather than the force acting on you at your interface with the elevator floor. From the requirement of static equilibrium, (we implicitly acknowledge that the 747 is moving with constant velocity), we conclude that lift force is equal to the weight, so to estimate the lift force we estimate the weight. Constructing this argument is half the problem. Now the other half: To estimate the weight we can guess... 100 tons? (More than 10 tons; we have heard of a 10 ton truck). But perhaps we can do better by slicing up the total weight, and try to estimate its ingredients. Passengers: How many passengers? I estimate 500. (More than 100, less than 1000). Then

$$weight\,/\,passenger \approx 250(pounds)$$

where I've thrown in an estimate of the weight of passenger luggage. This gives a contribution of

$$500 \times 250 = 125 \cdot 10^3 (pounds)$$

Structural weight: Let's focus on the weight of the fuselage and the weight of the wings... I imagine the fuselage to be a thin, circular cylinder– a tin can, an aluminum can, a *big* aluminum can. How big? How long, what diameter, what thickness? I will build up upon my estimate of number of passengers and my memory of seating arrangements in the big plane. I estimate 10 or 12 seats across, add another two seats for the two aisles, and taking seat width as two feet, I obtain a cylinder diameter of

$$14 \times 2 = 28 (feet)$$

I will round that up to 30 feet. Length: With 14 passengers per row and 500 passengers on board, and taking row pitch as 3 feet/row, I estimate total row length to be

$$3 \times 500/14 = 100 (feet)$$

That seems low. But other spaces must be accounted for. For example, galleys and restrooms, (another 40 feet?), and the pilot's compartment, (20 feet), the tail section (20 feet). Altogether then I estimate the length to be 180 feet. Thickness: I estimate 1/2 inch. Now I need a density, a weight density. Water is 62.4 pounds per cubic foot. The specific gravity of aluminum is what? I guess it to be 8. My estimate of fuselage weight is then the volume of the thin cylinder times the weight density. The volume is the circumference times the thickness – a good approximation when the thickness is small relative to the diameter – times the length. I obtain

$$\pi \times 30 \times 180 \times (1/2) \times (1ft/12in) \times 8 \times 62, 4 \approx 350 \cdot 10^3 (pounds)$$

Flooring and equipment and cosmetic structure will add more. I add another 20-25% and bring this up to

$$420 \cdot 10^3 (pounds)$$

The wings and engines come next: Here again we estimate the volume of wing material, now taking the wing as the equivalent of two thin sheets of aluminum. Wing length, tip to tip, as approximately the fuselage length, 180 feet. Wing breadth, or mean cord length we take as 20 feet, (the wing tapers as you move out from the fuselage to the tips of the wings) and the thickness again as .5 inches. The product of volume and weight density is then, noting that it take two such sheets to make the tip-to-tip wing

$$2 \times 180 \times 20 \times (1/2) \times (1/12) \times 8 \times 62, 4 \approx 150 \cdot 10^3 (pounds)$$

I throw in another 50,000 pounds for the motors and the tail (plus stabilizers) and so estimate the *total* structural weight to be

$$620 \cdot 10^3 (pounds)$$

Fuel: How much does the fuel weigh? The wings hold the fuel. I estimate the total volume enclosed by the wings to be the wing area times 1 foot. I take the density of fuel to be the same as water, 62.4 pounds per cubic foot. The total weight of the fuel is then estimated to be

$$2 \times 180 \times 20 \times 1 \times 62,4 \approx 450 \cdot 10^3 (pounds)$$

This looks too big. I can rationalize a smaller number citing the taper of the wing in its thickness as I move from the fuselage out to wing tip. I cut this estimate in half, not knowing anything more than that the tip volume must be near zero. So my fuel weight estimate is now

$$250 \cdot 10^3 (pounds)$$

All together then I estimate the lift force on a Boeing 747 at rush hour (fully loaded) to be

$$970 \cdot 10^3 \text{ pounds or approximately 500 tons.}$$

Is this estimate correct? Is it the right answer? Do I get an *A*? That depends upon the criteria used to differentiate right from wrong. Certainly we must allow for more than one numerical answer since there is *no* one numerical answer. If we admit a range, say of 20% either up or down, I may or may not pass. If we accept anything within a factor of 2, I am more confident, even willing to place a bet at 2 to 1 odds, that I am in the right. But go check it out: Jane's **All the World's Aircraft** will serve as a resource.

Is the method correct? The criteria here are more certain: In the first place, it is essential that I identify the lift force as the weight. Without this conceptual leap, without an abstraction of the plane as a particle, I am blocked at the start. This is a nontrivial and potentially argumentative step. More about that later.

Second, my method is more than a guess. It has a rationale, based upon a dissection of the question into pieces – passenger weight, structural weight, fuel weight – each of which in turn I might guess. But, again, I can do better: I dissect the passenger weight into a sum of individual weights. Here now I am on firmer ground, able to construct an estimate more easily and with confidence because an individual's weight is close at hand. So too with the structural and fuel weight; I reduce the question to simpler, more familiar terms and quantities. Fuel is like water in weight. The fuselage is a *big* aluminum can of football field dimensions. Here I have made a significant mistake in taking the specific gravity of aluminum as 8, which is that for steel. I ought to have halved that factor, better yet, taken it as 3. My total estimate changes but not by a factor of 2. The method remains correct.

Is this the *only* method, the only route to a rational estimate? No. A freshman thought of the weight of a school bus, fully loaded with forty passengers, and scaled up this piece. A graduate student estimated the lift force directly by considering the change in momentum of the airstream (free stream velocity equal to the cruising speed of the 747) as it went over the wing. There are alternative routes to follow in constructing an estimate; there is no unique single right method as there is no unique, single right number. This does *not* mean that there are *no* wrong methods and estimates or that some methods are not better than others.

Often you will not be able to develop a *feel* for the ingredients of an estimate or the behavior of a system, because of a disjunction between the scale of things in your experience and the scale of the problem at hand. If that be the case, try to break down the system into pieces of a more familiar scale, building an association with things you do have some feel for. More seriously, the dictates of the fundamental principles of static equilibrium might run counter to your expectations. If this is the case, stick with it. In time what at first seems counter-intuitive will become familiar.

Exercise 2.2

What do you need to know to determine the force in cables AB and BC?

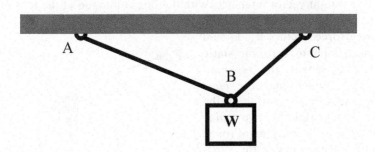

- You *need to know* that the cables support only tension. A *cable* is not able to support *compression* nor does it offer any resistance to *bending*. Webster's *New Collegiate Dictionary* notes the "great tensile strength" of a cable but says nothing about bending or compression. The cable's inability to support other than tension is critical to our understanding, our vocabulary. *Bending* itself will require definition... in time.

- You *need to know* the weight of the block.

- You *need to know* the angles AB and BC each make with the horizontal. (I will call them Θ_A and Θ_B). That's what you *need to know*.

- **You also must know how to isolate the system as a particle and you must know the laws of static equilibrium for an isolated particle.**

- Less apparent, you must presume that the force on the block due to gravity acts vertically downward — the convention in this textbook.

That is the answer to this *need to know* problem.

Now, you might ask me how do I know when to stop; how do I know when I don't need to know anymore? For example, how do I know I *don't* need to know the length of the cables or what material they are made from?

I know from solving this kind of problem before; you will learn when to stop in the same way – by working similar problems. I know too that the materials and lengths of the cables can be essential ingredients in the response to other questions about this simple system...e.g., if I were to ask *How much does the point B drop when the block is hooked up to the cables?* But that question wasn't asked.

Perhaps the best way to decide if you know enough is to try to solve the problem, to construct an answer to the question. Thus:

Exercise 2.3

Show that the forces in cable AB and BC are given by

$$F_C = W\cos\theta_A / \sin(\theta_A + \theta_C) \qquad \text{and} \qquad F_A = W\cos\theta_C / \sin(\theta_A + \theta_C)$$

We first isolate the system, making it a particle. Point *B*, where the *line of action* of the weight vector intersects with the lines of action of the tensions in the cables becomes our particle.

The three force vectors $\mathbf{F}_A$, $\mathbf{F}_C$ and $\mathbf{W}$ then must sum to zero for static equilibrium. Or again, the *resultant force on the isolated particle must vanish*. We meet this condition on the *vector sum* by insisting that two *scalar sums* – the sum of the horizontal (or *x*) components and the sum of the vertical (or *y*) components – vanish independently. For the sum of the *x* components we have, taking positive *x* as positive:

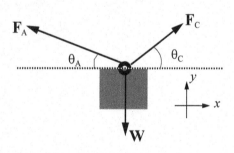

$$-F_A\cos\theta_A + F_C\cos\theta_C = 0$$

and for the sum of the *y* components, $F_A\sin\theta_A + F_C\sin\theta_C - W = 0$

A bit of conventional syntax is illustrated here in setting the sums to zero rather than doing otherwise, i.e., in the second equation, setting the sum of the two vertical components of the forces in the cables equal to the weight. Ignoring this apparently trivial convention can lead to disastrous results, at least early on in learning one's way in Engineering Mechanics. **The convention brings to the fore the necessity of isolating a particle before applying the equilibrium requirement.**

We see that what we *need to know* to determine the force in cable AB and in cable BC are the angles θ_A and θ_C and the weight of the block, W. These are the *givens*; the *magnitudes* of the two forces, F_A and F_C are our two *scalar unknowns*. We read the above then as two scalar equations in two scalar unknowns. We have reduced the problem ... *show that...* to a task in elementary algebra. To proceed requires a certain versatility in this more rarefied language.

There are various ways to proceed at this point. I can multiply the first equation by $\sin\theta_A$, the second by $\cos\theta_A$ and add the two to obtain

$$F_C \cdot (sin\theta_A \cdot cos\theta_C + sin\theta_C \cdot cos\theta_A) = W cos\theta_A$$

Making use of an appropriate trigonometric identity, we can write:[1]

$$F_C = W \cdot cos\theta_A / sin(\theta_A + \theta_C)$$

Similarly, we find:

$$F_A = W \cdot cos\theta_C / sin(\theta_A + \theta_C)$$

And thus we have shown what was asked to be shown. We have an answer, a unique answer in the sense that it is the only acceptable answer to the problem as stated, an answer that would merit full credit. But it is also an answer that has some depth, richness, a *thick* answer in that we can go beyond *show that* to *show and tell* and tease out of our result several interesting features.

- First, note that the derived equations are *dimensionally correct*. Both sides have the same units, that of force (or ML/T^2). In fact, we could easily obtain *non-dimensional* expressions for the cable tensions by dividing by the weight W. This *linear* relationship between the unknown forces in the cables and the applied load W will characterize most all of our discourse. It is a critical feature of our work in that it simplifies our task: If your boss asks you what will happen if some idiot accidentally doubles the weight hanging from the cables, you simply respond that the tensions in the cables will double. As always, there are caveats: We must assure ourselves first that the cables do not deform to any significant degree under double the weight. If they do then the angle θ_A and θ_C might change so a factor of 2.0 might not be quite exact. Of course if the cables break all bets are off.

---

1. This is an example of textbook rhetoric cryptically indicating a skipped step in the analysis. The author's presumption is that you, the reader, can easily recognize what's been left out. The problem is that it takes time, sometimes a long time, to figure out the missing step, certainly more time than it takes to read the sentence. If you are befuddled, an appropriate response then is to take some time out to verify the step.

- Second, note that the more vertically oriented of the two cables, the cable with its θ closer to a right angle, experiences the greater of the two tensions; we say it *carries the greatest load*.

- Third, note that the tension in the cables can be greater than the weight of the suspended body. The denominator $\sin(\theta_A + \theta_C)$ can become very small, approaching zero as the sum of the two angles approaches zero. The numerators, on the other hand, remain finite; $\cos\theta$ approaches 1.0 as θ approaches 0. Indeed, the maximum tension can become a factor of 10 or 100 or 1000... whatever you like... times the weight.

- Fourth, note the symmetry of the system when θ_A is set equal to θ_C. In this case the tensions in the two cables are equal, a result you might have guessed, or should have been able to claim, from looking at the figure with the angles set equal.[2]

- Fifth, if both angles approach a right angle, i.e., $\theta_A \to \pi/2$ *and* $\theta_C \to \pi/2$, we have the opportunity to use "L'Hospital's rule". In this case we have

$$F_C = F_A = \lim_{\theta \to \pi/2} (W\cos\theta / \sin2\theta) = \lim_{\theta \to \pi/2} (-W\sin\theta / 2\cos2\theta) = W/2$$

so each cable picks up half the weight.

Other observations are possible: What if one of the angles is negative? What if a bird sits on a telephone wire? Or we might consider the graphical representation of the three vectors in equilibrium, as in the following:

The figure below shows how you can proceed from knowing magnitude and direction of the weight vector and the directions of the lines of action of the forces in the two cables to full knowledge of the cable force vectors, i.e., their magnitudes as well as directions. The figure in the middle shows the directions of the lines of action of the two cable forces but the line of action of the force in cable AB, inclined at an angle θ_A to the horizontal, has been displaced downward.

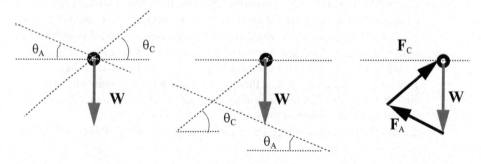

---

2. Buridan, a medieval scholar in Mechanics would have cited the *Principle of Sufficient Reason* in explaining how the forces must be equal for the symmetric configuration. There is no reason why one or the other cable tensions should be greater or less than the other. Buridan's ass, confronting two symmetrically placed bales of hay in front of his nose, starved to death. There was no sufficient reason to go left or right, so the story goes.

The figure at the far right shows the vectors summing to zero, that is, in vector notation we have:

$$\mathbf{F}_A + \mathbf{F}_C + \mathbf{W} = 0$$

While the two scalar equations previously derived in answering the *show that* challenge do not spring immediately from the figure, the relative magnitudes of the cable tensions are clearly shown in the figure at the far right.

- Graphical and algebraic, or analytic, constructions and readings of problems are complementary. Both should be pursued if possible; beyond two dimensions, however, graphical interpretations are difficult.

- Finally, note how the problem would have differed if the weight W and the angles θ_A and θ_C were stated as numbers, e.g., 100 pounds, 30 degrees, 60 degrees. The *setting-up* of the problem would have gone much the same but the solution would be thin soup indeed – two numbers, 50 pounds and 87.6 pounds, and that's about it; no opportunity for real thought, no occasion for learning about medieval scholars thoughts about sufficient reasoning, for conjecturing about birds on telegraph wires or applying L'Hospital's rule; it would be an exercise meriting little more than the crankings of a computer.

Resultant Force

We have used the phrase *resultant* in stating the requirement for static equilibrium of an isolated particle – the *resultant* of all forces acting on the isolated particle must vanish. Often we use *resultant* to mean the vector sum of a subset of forces acting on a particle or body, rather than the vector sum of *all* such forces. For example, we can say "the resultant of the two cable tensions, $\mathbf{F}_A$ and $\mathbf{F}_C$ acting at point B in Exercise 2.3 is the force vector $-\mathbf{W}$". The resultant is constructed using the so called *parallelogram law for vector addition* as illustrated in the figure. $\mathbf{F}_A + \mathbf{F}_C$ can then be read as $-\mathbf{W}$, the vector equal in magnitude to the weight vector $\mathbf{W}$ but oppositely directed.

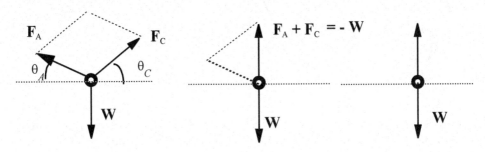

We can also speak of a single force vector as being the *resultant* of its components, usually its three mutually perpendicular (or *orthogonal*) rectangular, cartesian components. In the figure, the vector **F** has components $\mathbf{F}_x$, $\mathbf{F}_y$ and $\mathbf{F}_z$.

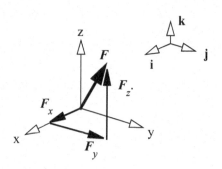

The latter three mutually perpendicular vectors are usefully written as the product of a scalar magnitude and a unit vector indicating the direction of the vector component. The three unit vectors are often indicated by **i**, **j** and **k** and that is the convention we will follow in this text.

The vector resultant or sum can be written out as

$$\mathbf{F} = \mathbf{F}_x + \mathbf{F}_y + \mathbf{F}_z$$

or, equivalently

$$\mathbf{F} = F_x\mathbf{i} + F_y\mathbf{j} + F_z\mathbf{k}$$

Note the convention for designating a vector quantity using boldface. This is the convention we follow in the text. In lecture it is difficult to write with chalk in boldface. It is also difficult for you to do so on your homework and exams. In these instances we will use the convention of placing a bar (or twiddle) over or under the letter to indicate it is a vector quantity.

Exercise 2.4

For each of the three force systems shown below, estimate the magnitude of the resultant, of F_1 and F_2. What is the direction of the resultant in each case?

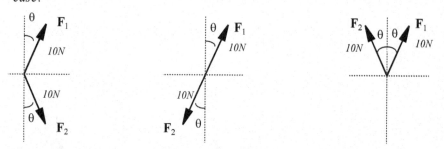

For the force system at the left, the vertical component of the resultant force must be zero since the vertical components of the two forces are equal but one is positive (upward) and the other negative (downward). The symmetry with respect to the horizontal also leads to the result that the horizontal component of the resultant will be the sum of the two (equal) horizontal components of the two forces. I estimate this sum to be 8 *Newtons* since the angle θ looks to be something less than 30°. The resultant is directed horizontally to the right.

For the system in the middle, the resultant is zero since the two forces are equal but oppositely directed. If we were concerned with the static equilibrium of these three systems, only this system would be in equilibrium.

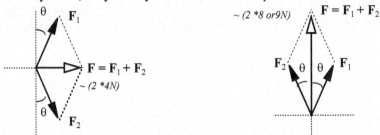

For the system at the far right, with the two forces symmetric with respect to the vertical, we now have that the horizontal component of the resultant force must be zero while the vertical component of the resultant will be the sum of the two (equal) vertical components of the two forces. I estimate this sum to be 16 ~ 18 *Newtons*. The resultant is directed vertically upward as shown above.

Friction Force

Often the greatest challenges in applying the requirements of static equilibrium to useful purpose is isolating the particle (or later the body) and showing on your isolation, your *free-body diagram*, all the forces (and later all the torques as well) that act. These include reaction forces as well as forces applied like the weight due to gravity. Imagining and drawing these forces takes a certain facility in the creative use of the language of Engineering Mechanics, in particular, a facility with the characteristics of different kinds of forces. One of the kinds you will be responsible for reading out of a problem statement and writing into your free-body diagram is the *force due to friction*.

Friction is tricky because sometimes it can be anything it needs to be; it's direction as well as magnitude have a chameleon quality, taking on the colors that best meet the requirements of static equilibrium. But it can only be so big; it's magnitude is limited. And when things begin to move and slide, it's something else again. Friction is even more complicated in that its magnitude depends upon the surface materials which are in contact at the interface you have constructed in your free-body diagram but *not* upon the area in contact. This means that you have to go to a table in a reference book, ask a classmate, or call up a supplier, to obtain an appropriate value for the *coefficient of friction*.

Consider the following illustration of the practical implications of friction and the laws of static equilibrium: I know from experience that my back goes out if I pull on a cable angling up from the ground with a certain force, approximately fifty pounds. Will my back go out if I try to drag the heavy block shown by pulling on the cable *AB*?

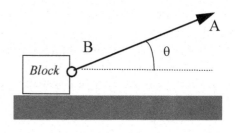

To find the force I must exert, through the cable, onto the block in order to slide it to the right I will isolate the block and apply the laws of static equilibrium. Now this, at first reading, might appear a contradiction: How can the block be sliding to the right and in static equilibrium at the same time?

Two responses are possible:

- If the block is sliding to the right at constant velocity, then the laws of static equilibrium still apply – as when you ascended in an elevator at constant velocity. There is no contradiction.

- If the block does not move, but is *just about to move*, then the laws of static equilibrium still apply and again there is no contradiction. In this case we say we are at the point of *impending motion*; the smallest increment in the force with which I pull on the cable will start the block sliding to the right.

It is the second case that I will analyze, that of *impending motion*. It is this case that will most likely throw my back out.

I first isolate the block as a particle, showing all the applied and reaction forces acting upon it. The weight, **W**, and the force with which I pull the cable *AB*— I will call it $\mathbf{F}_A$— are the applied forces.

The reaction forces include the force of the ground pushing up on the block, **N**, what is called a *normal force*, and the friction force, $\mathbf{F}_f$, acting

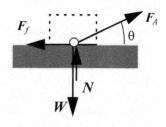

parallel to the plane of contact of the block with the ground, tending to resist motion to the right, hence acting on the block to the left. For static equilibrium the resultant force on the particle must vanish. Again, this is equivalent to demanding that the sum of the horizontal components and the sum of the vertical components vanish independently.

Thus: $F_A \cos\theta - F_f = 0$ and $F_A \sin\theta + N - W = 0$

These are two scalar equations. But look, there are at least three unknowns – F_f, F_A and N and θ. Even if θ is given, say 60°, we are still in a fix since there remains one more unknown over the number of independent equations available. Now it's not that we can't find a solution; indeed we can find any number of sets of the three unknowns that serve: Just pick a value for any one of the three – F_A, F_f, or N as some fraction of W, say W/2, and use the equilibrium equations to solve for the remaining two. The problem is we can not find a *unique* solution. We say that system is under determined, or indeterminate.

This is where *impending motion* comes to our rescue. We add the condition that we are at the point of impending motion. At impending motion the frictional force is related to the normal force by

$$F_f = \mu \cdot N$$

where μ is called *the coefficient of static friction.*

Note that μ is a *dimensionless* quantity since both the normal force N and its associated friction force F_f have the same dimensions, that of force. The particular value of the coefficient of static friction depends upon the materials in contact at the interface where the friction force acts. Another coefficient of friction is defined to cover the case of sliding at constant velocity. It is labeled the coefficient of sliding friction. It too depends upon the character of the materials in contact; its value is nominally less than μ.

This is the third equation that allows us to estimate the force that will throw my back out. In fact, solving the three equations we find:

$$F_A = \mu \cdot W / (cos\theta + \mu \cdot sin\theta)$$

From this expression I can estimate the weight of the block I might be able to move by pulling on the cable *AB*. For example, if $\theta = 60^o$ and I take $\mu = 0.25$, as an estimate for sliding blocks along the ground, then setting $F_A = 50$ *lb.* — an estimate of the maximum force I can exert without disastrous results — and solving for W, I find from the above

$$W = 143.3 \text{ lb.}$$

The friction force at impending motion is in this case, from the first equation of equilibrium,

$$F_f = F_A cos\theta = 50 \cdot cos 60^o = 25 \text{ lb.}$$

If I pull with a force *less than* fifty pounds, say twenty pounds, still at an angle of 60°, on a block weighing 143.3 pounds or more, the block will not budge. The friction force F_f is just what it needs to be to satisfy equilibrium, namely $F_f = 20$ cos $60^o = 10$ *lb*. This is what was meant by the statement " ... can be anything it has to be." The block in this case does not move, nor is it just about to move. As I increase the force with which I pull, say from twenty to fifty pounds, the frictional force increases *proportionally* from ten to twenty-five pounds, at which point the block begins to move and we leave the land of Static Equilibrium. That's okay; I know now what weight block I can expect to be able to drag along the ground without injury. I need go no further.

But wait! What, you say, if I *push* instead of pull the block? Won't pushing be easier on my back? You have a point: I will now analyze the situation given that the *AB* is no longer a cable which I pull but signifies my arms pushing. In this we keep θ equal to 60°.

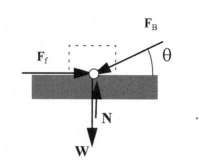

At first glance, you might be tempted as I was, when I was a youth with a good back but little facility in speaking Engineering Mechanics, to simply change the sign of F_A in the equilibrium equations and let it go at that: Solving would simply change the sign in our final expression for F_A in terms of W. But that will not do.

Friction is trickier: **Friction always acts in a direction resisting the impending motion**. Here is another way it changes its colors to suit the context. No, I can't get away so simply; I must redraw my free-body diagram carefully showing the new directions of the forces F_A *and* F_f. In this I will label the force I apply, F_B.

Equilibrium now gives, taking horizontal components to be positive when directed to the right and vertical components positive when directed upward:

$$-F_B cos\theta + F_f = 0 \quad \text{and} \quad -F_B sin\theta + N - W = 0$$

These two, again supplemented by the relationship between the friction force and the normal force, namely

$$F_f = \mu \cdot N$$

now yield

$$F_B = \mu \cdot W /(cos\theta - \mu \cdot sin\theta)$$

This is the force I must push with in order to just start the block sliding to the left – a state of impending motion. If I push with a force less than this, the block will not budge, the friction force is whatever the first equilibrium equation says it has to be. What force must I push with in order to move a block of weight $W =$ 143.3 *lb*? Again, with θ = 60°, the above expression gives F_B = 126.3 lb.

Note well the result! I must push with more than twice the force I must pull with in order to move the block! There is no mystery here. The reasons for this are all contained in the equations of equilibrium and the rules we have laid out which govern the magnitude and direction of the force due to friction. It is the latter that adds so much spice to our story. Note, I might go on and construct a story about how pushing down at an angle adds to the normal force of reaction which in turn implies that the frictional force resisting motion at the point of impending motion will increase. The bottom line is that it takes more force to push and start the block moving than to pull and do the same.

Exercise 2.5

Professor X, well known for his lecturing theatrics, has thought of an innovative way to introduce his students to the concept of friction, in particular to the notion of impending motion. His scheme is as follows: He will place a chair upon the top of the table, which is always there at the front of the lecture hall, and ask for a volunteer from the class to mount the table and sit down in the chair. Other volunteers will then be instructed to slowly raise the end of the table. Students in the front row will be asked to estimate the angle φ as it slowly increases and to make a note of the value when the chair, with the student on board, begins to slide down the table surface, now a ramp.

Unfortunately, instead of sliding, the chair tips, the student lurches forward, fractures his right arm in attempting to cushion his fall, gets an A in the course, and sues the University. Re*construct* what Professor *X* was attempting to demonstrate and the probable cause of failure of the demonstration?

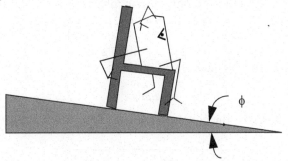

I begin by drawing a free-body diagram, isolating the student and the chair together – all that which will slide down the table top when tipped up – as a particle. The weight W in the figure below is the weight of the student and the chair. The reaction force at the interface of the chair with the table is represented by two perpendicular components, the normal force N and the friction force F_f. We now require that the resultant force on the particle vanish.

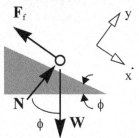

In this, we choose the *x-y* axes shown as a reference frame. We make this choice to simplify our analysis. Only the weight vector W has both *x* and *y* components. Make a mental note of this way of crafting in setting up a problem. It is a bit of nuance of language use that can help you express your thoughts more efficiently than otherwise and yield a rich return, for example, on a quiz when time is precious. Equilibrium in the *x* direction, positive down the plane, then requires

$$-F_f + W \cdot sin\phi = 0$$

while equilibrium in the *y* direction yields

$$N - W \cdot cos\phi = 0$$

I manipulate these to obtain

$$F_f = N tan\phi \qquad \text{where} \qquad N = W \cdot cos\phi$$

Now we know from the previous friction problem we analyzed that the friction force can only get so large *relative to the normal force* before motion will ensue. For the problem at hand, once the ratio of friction force to normal force reaches a value equal to the coefficient of static friction, μ, appropriate for the chair's leg tips interfacing with the material of the table's top, motion of the "particle" down the plane will follow. We can state this condition as an *inequality*. The student and chair will not slide down the plane as long as

$$F_f = N \cdot tan\phi < \mu \cdot N$$

This immediately yields the conclusion that as long as the angle ϕ is less than a certain value, namely if

$$\boxed{tan\phi < \mu}$$

the particle will not move. Note that, on the basis of this *one-to-one* relationship, we could define the condition of impending motion between two materials in terms of the value of the angle ϕ as easily as in terms of μ. For this reason, ϕ is sometimes called the *friction angle*. For example, if $\mu = 0.25$ then the angle at which the chair and student will begin to slide is $\phi = 14^o$ Note, too, that our result is independent of the weights of the student and the chair. *All* students should begin to slide down the plane at the same angle. This was to be a central point in Professor *X*'s demonstration: He planned to have a variety of students take a slide down the table top. Unfortunately the tallest person in class volunteered to go first.

Why did the demonstration fail?   It   failed because Professor X saw a *particle* where he should have envisioned an *extended body*. The figure at the right is an adaptation of a sketch drawn by a student in the front row just at the instant before the student and chair tipped forward. Note that the *line of action* of the weight vector of the chair-student combination, which I have added to the student's sketch, passes

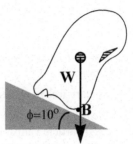

through the point of contact of the front legs of the chair with the table top, point B. Note, too, that the angle ϕ is less than the friction angle, less than the value at which the chair would begin to slide.

In the next instant, as the students charged with lifting the left end of the table did as they were told and raised their end up an inch, the line of action of W fell forward of point B and the accident ensued.

When is a particle a particle?

The question perhaps is better phrased as "When is a body a particle?" The last exercise brings forcibly home how you can go wrong if you mistakenly read a particle where you ought to imagine something of more substance. We have here a failure in *modeling*.

Modeling is a process that requires you to represent "reality" in the language of Engineering Mechanics, to see in the world (or in the text in front of you) the conceptual ingredients of force, now of torque or moment and how the laws of static equilibrium and subsidiary relations like those that describe the action of a force due to friction, are to apply. It was Professor X's failure to see the tipping moment about point B that led to his, rather the student's, downfall.

Modeling failures are common, like the cold. And there is no easy fix nor medicine to prescribe that will ensure 100% success in modeling. One thing is essential, at least here at the start: You must draw an isolation, a free-body diagram, as a first, critical step in your response to a problem. That until now, has meant, not just drawing a point on a clean sheet of paper – anyone can do that – but that you imagine all the force vectors acting on the particle and draw them too on your sheet of paper.

This requires some thought. You must imagine; you must take risks; you must conjecture and test out your conjecture. In this you have available the beginnings of a vocabulary and some grammar to help you construct an appropriate isolation of, at least, a particle:

- gravity acts downward;
- friction force acts to resist motion;
- the normal force acts perpendicular to the plane of contact;
- a cable can only sustain a tensile force;

- to every action there is an equal and opposite reaction.

To these notions we can add:

- it doesn't matter where you show a force vector acting along its *line-of-action*;[3]
- you are free to choose the orientation of a reference coordinate system;
- the requirement that the resultant force of *all* the forces acting on an *isolated* particle vanish is equivalent to requiring that the sums of the usually orthogonal, scalar components of all the forces vanish.

Knowing all of this, there remains ample room for error and going astray. Professor *X*'s free-body diagram and analysis of a particle were well done. The failing was in the field of view right at the outset; Modeling a student in a chair on an inclined plane as a particle was wrong from the start.

Now this really makes life difficult since, for some purposes, the chair and student might be successfully modeled as a particle, e.g., if the coefficient of friction is sufficiently small such that I need not worry about the tipping forward, (see problem 2.12), while at other times this will not do. Or consider the block I pull along the ground in Exercise 2.4: I successfully modeled the block as a particle there. Note how, at this point, you might now conjecture a scenario in which I could *not* claim success, for example, if the geometry were such that the block would lift off the ground before sliding. Or consider the airplane of Exercise 2.1. If I am interested in the resultant lift force, I can get away with modeling the football-field size machine as a particle; on the other hand, if I were responsible for defining how to set the flaps to maintain a specified attitude of the craft, I would have to take the airplane as an extended body and worry about the distribution of the lift force over the wings and the horizontal stabilizer.

We conclude that the chair and student, indeed all things of the world of Engineering Mechanics, do not appear in the world with labels that say "I am a particle" or "I am not a particle". No, it is *you* who must provide the labels, read the situation, then articulate and compose an abstract representation or *model* that will serve. In short, something might be a particle or it might be an extended body depending upon your interests, what questions you raise, or are raised by others for you to answer.

Now, in most texts whether something is a particle or a body can be easily imputed from the context of the problem. You expect to find only particles in a chapter on "Static Equilibrium of a Particle". On the other hand, if an object is dimensioned, i.e., length, width and height are given on the figure, you can be quite sure that you're meant to see an extended body. This is an usually unstated rule of textbook writing – authors provide all the information required to solve the problem and *no* more. To provide more, or less, than what's required is considered, if not a dirty trick, not in good form. I will often violate this norm. Engi-

---

3. This is true as long as we are not concerned with what goes on within the boundary we have drawn enclosing our free body.

neers, in their work, must deal with situations in which there is an excess of information while, at other times, situations in which there is insufficient information and conjecture and estimation is necessary. It's best you learn straight off a bit more about the real world than the traditional text allows.

2.2   Concept of Moment

Force is not enough. You know from your studies in physics of the dynamics of bodies other than particles, that you must speak about their rotation as well as translation through space; about how they twist and turn.

Equilibrium of a Body

We turn, then, to consider what we can say about forces, applied and reactive, when confronted with a body that cannot be seen as a particle but must be taken as having finite dimensions, as an *extended* body. Crucial to our progress will be the concept of *moment* or *torque* which can be interpreted as the *turning effect of a force*.

We start again with a block on the ground. Instead of pushing or pulling, we explore what we can do with a lever. In particular we pose, as did Galileo (who also had a bad back),

Exercise 2.6

Estimate the magnitude of the force I must exert with my foot pressing down at B to just lift the end of the block at A up off the ground?

We isolate the system, this time as an extended body, showing all the applied and reaction forces acting on the system. The applied forces are the weight acting downward along a vertical line of action passing through the *c.g.*, the *center of*

gravity of the block, and the force of my foot acting downward along a line of action through the point *B* at the right end of the lever *AB*.

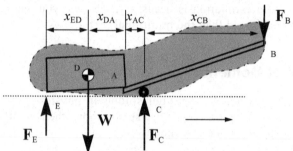

The reaction forces are two: (1) the force of the ground acting up on the left end of the block at *E* and (2), the force of the ground acting up through the pivot at *C* upon the lever *AB*. Our quest is to determine the magnitude of the (vertical) force we must apply at *B* in order to *just lift* the end *A* off the ground. We start by applying our known requirement for static equilibrium – **for a body at rest, or moving with constant velocity, the resultant force acting on the** *isolated body* **must be zero, must vanish**. We have, taking up as positive,

$$F_E - W + F_C - F_B = 0$$

We read this as one (scalar) equation with three unknowns, the applied force F_B and the two reactions F_E and F_C. Clearly we need to say something more. That "more" is contained in the following equilibrium requirement for an extended body – **for a body at rest, or moving with constant velocity, the resultant moment of all forces acting on the** *isolated body* **must be zero, must vanish**. I will find the resultant *moment* or *torque* of all the forces *about the left-most point E*. I will take as positive, a torque which tends to rotate the extended body of block and lever – all that lies within the dotted envelope – clockwise. For example, the moment about point *E* of the reaction force F_C is negative since it tends to rotate the system counter-clockwise about the reference point *E*. Its value is given by $(x_{ED}+x_{DA}+x_{AC})\, F_C$, the product of the force F_C and the perpendicular distance from the point *E* to the *line of action* of the force F_C.

The *resultant moment* of all the forces acting on the isolated system is

$$x_{ED} \cdot W - (x_{ED} + x_{DA} + x_{AC}) \cdot F_C + (x_{ED} + x_{DA} + x_{AC} + x_{CB}) \cdot F_B = 0$$

We may read this equation as a second scalar equation in terms of the three unknown force quantities *if* we take the *x*'s, the distance measures, as known. We might, at this point, estimate the distances: the block length looks to be about one meter.[4] Then, from the figure, estimate the other lengths by measuring their magnitudes relative to the length of the block. I will not do this. Instead, for reasons that will become evident, I will *not* state the block length but simply label it L then figure the x's in terms of L.

---

4. A better estimate might be obtained if the reader could identify the shrub at the left of the block. But that's beyond the scope of the course.

My estimates for the lengths are then:

$$x_{ED} = x_{DA} = L/2 \qquad x_{AC} = L/5 \qquad x_{BC} = 7L/5$$

With these, my *equation of moment equilibrium* becomes

$$(L/2) \cdot W - (6L/5) \cdot F_C + (13L/5) \cdot F_B = 0$$

Now make note of one feature. L, the length of the block is a common factor; it may be extracted from each term, then "cancelled out" of the equation. We are left with

$$(W/2) - (6/5) \cdot F_C + (13/5) \cdot F_B = 0$$

Where do we stand now? We have two equations but still three unknowns. We are algebraically speaking "up a creek" if our objective is to find some one, useful measure of the force we must exert at B to just lift the block of weight W off the ground at the end A. Again, it's not that we cannot produce a solution for F_B in terms of W; the problem is we can construct many solutions, too many solutions, indeed, an infinite number of solutions. It appears that the problem is indeterminate.

In the next chapter we are going to encounter problems where satisfying the equilibrium requirements, while necessary, is not sufficient to fixing a solution to a problem in Engineering Mechanics. There we will turn and consider another vital phenomenon - the deformation of bodies. At first glance we might conclude that the problem before us now is of this type, is *statically indeterminate*. That is **not** the case. Watch!

I will dissect my extended body, isolating a portion of it, namely the block alone. My *free-body diagram* is as follows:

I have constructed a new force F_A, an *internal force*, which, from the point of view of the block, is the force exerted by the end of the lever at A upon the block. Now **this** extended body, **this** subsystem is also in static equilibrium. Hence I can write

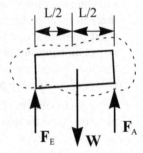

$$F_E - W + F_A = 0$$

ensuring *force equilibrium* and $\qquad W \cdot (L/2) - F_A \cdot L = 0 \qquad$ ensuring *moment equilibrium about point E*. The second equation gives us directly $\quad F_A = W/2$

while the first then yields $\quad F_E = F_A = W/2 \quad$ which we might have concluded from the symmetry of our free-body diagram[5]. My next move is to construct yet another isolated body, this time of the lever alone.

5. Perceiving this symmetry depends upon knowing about the requirement of moment equilibrium of an isolated body so it's a bit unfair to suggest you might have been able to "see" this symmetry without this knowledge.

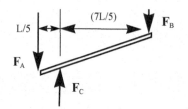

Note that the force acting on the lever due to the block is the *equal and opposite internal reaction* F_A whose magnitude we now know. **This is an essential observation**. Now **this** extended body, **this** subsystem is also in static equilibrium. Hence I can write

$$-W/2 + F_C - F_B = 0$$

ensuring *force equilibrium* and

$$-(L/5) \cdot W/2 + ((7L)/5) \cdot F_B = 0$$

ensuring *moment equilibrium about point C*. Note the repetition in language here and with that of the analysis of the free-body diagram of the block alone. Indeed, once we have constructed the abstract representation, the free-body diagram, the subject matter becomes somewhat boring and repetitive, machine-like. From the second equation, that of moment equilibrium, I find $F_B = W/14$ which is a significant *mechanical advantage*.

- Observe that if I had *taken moments* about point *A* shown in the free-body diagram of the lever, I would have obtained a *different* equation expressing moment equilibrium namely, $-(L/5) \cdot F_C + ((8L)/5) \cdot F_B = 0$

- but I would obtain the same result, the same answer. However, I would have to make use of the equation of force equilibrium together with this last equation of moment equilibrium to get to the answer. This feature of this particular problem may be generalized, to wit: **It doesn't matter what point in space you choose as a reference point when you construct an equation of moment equilibrium.** This is powerful knowledge that may dramatically increase your productivity for often, by judicious choice of a reference point, you can simplify your analysis.

- Observe too that in all three isolations the forces were read as planar and parallel, that is their lines of action were drawn in a single plane and parallel to the vertical. In each of the three cases, for each isolation, we wrote out two independent, scalar equations; one expressed force equilibrium in the vertical direction, the other moment equilibrium about some reference point. Now I could have, instead of force equilibrium, applied moment equilibrium again, about some *other* reference point. For example, for the lever, the last isolation diagram constructed, if I take moment equilibrium about the left end, this, together with the consequence of moment equilibrium about point *C*, namely $F_B = W/14$ produces the same result for the reaction at *C*. Check it out.

- Observe that at a point early on in our analysis we might have concluded that we had insufficient information to do the problem. But, by breaking down the problem into two other problems we found our way to a solution.

- Observe, finally, that, after having analyzed the block as an isolated sub-system and obtained the reaction force, F_E, we could have gone directly back to the original three equations of equilibrium and solved for the remaining two unknowns, F_C and F_B. Once again we note that there are alternative paths to a solution. Some paths are more direct than others; some are more enlightening than others, but they all should lead to the *same* solution if the question is well-posed.

We begin to see now the more subtle aspects of applying the requirements of static equilibrium to useful purpose: Effective use of this new language will require us to make choices – choices of reference points for taking moments, selection of subsystems to analyze when one free-body diagram won't yield all we need to know – and requires a familiarity with different renderings of force and moment equilibrium. There is no unique, cook-book, 100% sure method to solving problems, even statics problems, in Engineering Mechanics.

Different Kinds of Systems of Forces

The requirements of force equilibrium and moment equilibrium are two *vector* equations. We can write them as:

$$\sum \mathbf{F_i} = 0 \qquad \sum \mathbf{M_i} = 0$$

In these two **vector** equations, the summation is to be carried out over all forces and moments acting on an isolated body— the i ranges over one to N forces say. We can interpret the first as the resultant of all externally applied forces and the second as the resultant moment of all the forces (and other, *concentrated moments or couples,* yet to be defined) acting on the body.

The resultant force on a particle or body and the resultant moment are both vector quantities; each has a magnitude and a direction which must be specified to fully know the nature of the beast. Each vector resultant has three (3) scalar components in three-dimensional space so each vector equation is equivalent to three *independent* scalar equations. From this we conclude:

- **There are at most six (6) independent scalar equations available (which must be satisfied) to ensure static equilibrium of an isolated body. For a particle, there are at most three (3) independent scalar equations available.**

If you look back over the exercises we have worked in the preceding sections of this text you will note: – estimating the lift force on an aircraft required citing a single scalar equation; in pulling and pushing the block along the ground, with the block taken as a particle, we made use of two scalar equations; so too, our analysis of a particle sliding down a plane required the use of two scalar equations of equi-librium. Nowhere did we need three scalar equations of equilibrium. The reason? All force vectors in each of these particle problems lay in the plane of the page hence each had but two scalar components, two x,y or, in some cases, horizontal, vertical components. Likewise the resultant force shows but two scalar components. Force equilibrium is equivalent to setting the sum of the x components and the sum of the y components to zero. Alternatively we could say that force equilibrium in the direction perpendicular to the plane of the page is *identically satisfied*; 0 = 0; since there are no components in this direction.

In our analysis of an extended body, the block with lever applied, we had six scalar equations available, at most. Yet in each of the three isolations we constructed we wrote but two independent scalar equations and that was sufficient to our purpose. How do we explain our success; what about the other four scalar equations? They must be satisfied too.

First note that again, all force vectors lie in the plane of the page. Not only that, but their lines of action are all parallel, parallel to the vertical. Hence force equilibrium in all but the vertical direction is satisfied. That takes care of two of the four.

Second, since the force vectors all lie in a single plane, they can only produce a turning effect, a torque or a moment, about an axis perpendicular to that plane. Thus moments about the axes lying *in the plane*, the x,y axes, will be identically zero. That takes care of the remaining two scalar equations not used.

From all of these observations we can boldly state:

- **If the lines of action of all forces acting on a *particle* lie in a common plane, there are at most two independent, scalar, equilibrium equations available.**

- **If the lines of action of all forces acting on an *extended body* are all parallel and lie in a common plane there are at most two independent, scalar, equilibrium equations available.**

- **If the lines of action of all forces acting on an *extended body* all lie in a common plane there are at most three independent, scalar, equilibrium equations available.**

Note well, however, that these are statements about the *maximum number* of independent equations available to us in particular contexts. They do *not* say that so many must derive from moment equilibrium and so many from force equilibrium. We have seen how, in the analysis of the lever used in lifting the end of Galileo's block up off the ground, we were able to apply moment equilibrium twice to obtain a different looking, but equivalent, system of two linearly independent equations – different from the two obtained applying moment equilibrium once together with force equilibrium in the vertical direction.

Similarly for the block sliding, as a particle, down the plane we might have oriented our x,y axes along the horizontal and vertical in which case the two scalar equations of force equilibrium would appear, for the horizontal direction,

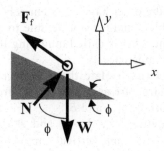

$$-F_f \cdot \cos\phi + N \cdot \sin\phi = 0$$

and, for the vertical direction,

$$N \cdot \cos\phi + F_f \cdot \sin\phi - W = 0$$

These two are equivalent to the two-force equilibrium equations we previously derived; they yield the same solution; they are synonymous; the two sets have identical meaning.

The requirements of force and moment equilibrium have further implications for particular systems of forces and moments. We use them to define *two force members* and *three-force members*.

Exercise 2.7

What do you need to know in order to determine the reaction forces at points A and B of the structure shown below?

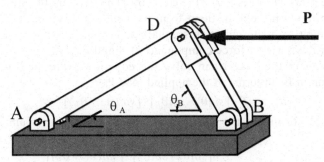

In that this problem, at least with regard to its geometry, looks like the block hanging from two cables of Exercise 2.2, we might begin by mimicking what we discovered we needed to know there:

- You need to know the force applied P, just as we needed to know the weight of the block in Exercise 2.2.

- You need to know the angles θ_A and θ_B

The other item needed-to-know doesn't apply since members AD and BD are not cables. To go further we will now try to solve the problem. If we are able to solve the problem we surely then should be able to say what we needed to know.

On the other hand, it is entirely possible that a neophyte might be able to solve a problem and yet not be able to articulate what they needed to know to get to that point. This is called muddling through and ought not to be condoned as evidence of competence.

I first isolate the system, cutting the structure as a whole away from its moorings at A and B.

Letting A_x, A_y, B_x and B_y be the horizontal and vertical components of the reaction forces at A and B, force equilibrium will be assured if

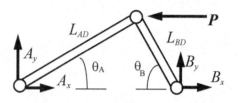

$$A_x + B_x - P = 0$$

and

$$A_y + B_y = 0$$

Our third equilibrium equation for this planar system of forces acting on an extended body may be obtained by *taking moments about point A*[6].

We have

$$(L_{AD} \cdot cos\theta_A + L_{BD} \cdot cos\theta_B) \cdot B_y - P \cdot L_{AD} \cdot sin\theta_A = 0$$

In this, I have taken clockwise as positive. Observe that the two lengths that appear in this equation are related by

$$L_{AD} \cdot sin\theta_A = L_{BD} \cdot sin\theta_B$$

At first glance it would appear that I must know one of these lengths in order to solve the problem. This is *not* the case since the two lengths are *linearly related* and, after eliminating one of them from moment equilibrium, the other will be a factor common to all terms in this equation, hence, will not appear in the final solution for the reaction force components at A and B.

Proceeding in this way I can solve the equation expressing moment equilibrium and obtain B_y in terms of the applied load P.

$$B_y = -P \cdot [sin\theta_B \cdot sin\theta_B] / [sin(\theta_A + \theta_B)]$$

Force equilibrium in the y direction, the second equilibrium equation, then gives

$$A_y = -B_y = P \cdot [sin\theta_B \cdot sin\theta_B] / [sin(\theta_A + \theta_B)]$$

So far so good.... but then again, that's as far as we can go; we are truly up a creek. There is no way we can find unique expressions for A_x and B_x in terms of P, θ_A, and θ_B without further information to supplement the first equilibrium equation. But watch!

---

6. This is a bit of common dialect, equivalent to stating that the sum of the moments of all forces with respect to A will be set to zero in what follows. Note too how, in choosing A as a reference point for moments, the components of the reaction force at A do not enter into the equation of moment equilibrium.

I isolate a subsystem, member *AD*. The circles at the ends of the member are to be read as *frictionless pins*. That means they can support *no* moment or torque locally, at the end points.

These circles act like a ring of ball bearings, offering no resistance to rotation to the member about either end. This implies that only a force can act at the ends. These might appear as at the right. Now we apply equilibrium to this isolated, extended body $\mathbf{F}_A + \mathbf{F}_D = 0$.

This implies that the two internal forces must be equal and opposite. Our isolation diagram might then appear as on the left, below

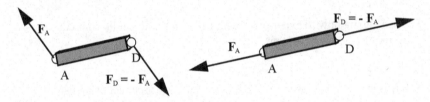

Force equilibrium is now satisfied but surely moment equilibrium is not. The two forces as shown produce an unbalanced *couple*. For this to vanish we require that the two lines of action of the two forces be coincident, co-linear and our isolation diagram must appear as on the right.

We can go no further. Yet we have a very important statement to make:

- **For a two-force member, an isolated body with but two forces acting upon it, the two forces must be equal, opposite, and colinear.**

For the straight member *AD*, this means that the lines of action of the forces acting at the ends of the member lie along the member. We say that *the two-force (straight) member is in tension or compression.*

Knowing that the force at *A* acting on the member *AD* acts along the member, we now know that the direction of the reaction force at *A* because the reaction force at *A* is the same force. This follows from isolating the support at *A* .

Here we show the reaction force at *A* with its components A_x and A_y now drawn such that $A_y/A_x = tan\theta_A$

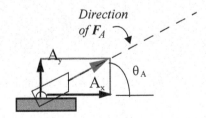

This last equation, together with the first equilibrium equation and our previously obtained expression for A_y, enables us to solve for both A_x and B_x. And that completes the exercise.

At this point, the reader may feel cheated. Why not go on and find the reaction forces at the points A and B? No. I have solved the *need to know problem* as it was posed. I have constructed a well posed problem; I can farm it out to some subordinate to carry through the solution, confident that anyone with a basic knowledge of algebra can take it from here. The "heavy lifting" is over. So too, you must learn to have confidence in your ability to set-up a well posed problem and to delegate the responsibility for crank-turning to others, even to computational machinery. *However*, you, of course, remain responsible for evaluating the worth of what comes back to you.

In conclusion, in addition to knowing the load P and the angles θ_A and θ_B, in order to solve the problem we need to know:

- How to read a circle as a frictionless pin and what we mean by the phrase *two-force member.*

- Also, again, how to experiment with isolations and equilibrium considerations of pieces of a system.

Some final observations:

- Say this structure was the work of some baroque architect or industrial designer who insisted that member *BD* have the form shown immediately below. While there may be legitimate reasons for installing a curved member connecting *B* to *D*, — e.g., you need to provide adequate clearance for what is inside the structure, this creates no problem. The reactions at *A* and *B* (see below) will be the same as before (assuming frictionless pins). What *will* change is the mode in which the baroque, now curved, member *BD* carries the load. It is no longer just in compression. It is subject to *bending*, a topic we consider in a section below[7].

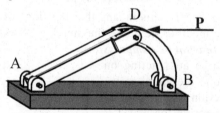

- Returning to our original structure with straight members, the same sort of analysis of an isolation of member *BD* alone would lead us to conclude that the direction of the reaction at *B* must be up along the member *BD*. Is

---

7. Independent of such motives, some will argue that "form follows function"; the baroque for baroque's sake is not only functionally frivolous but ugly.

this the case? That is, do the values obtained for B_x and B_y obtained above satisfy the relationship $\tan\theta_B = -B_y/B_x$? Check it out!

Exercise 2.8

Show that the force F required to just start the lawn roller, of radius R and weight W, moving up over the ledge of height h is given by

$$F/W = \tan\phi \qquad \text{where} \qquad \cos\phi = 1 - (h/R).$$

We, as always, start by isolating the system – the roller – showing all the

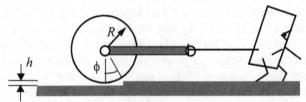

forces acting upon it. These include the weight acting downward through the center of the roller; the horizontal force applied along the handle by the child laborer (note the frictionless pin fastening the handle to the roller); and the components of the reaction force at the bump, shown as N_x and N_y.

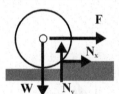

Note the implications of the phrase *just start...* This is to be read as meaning there is no, or insignificant, contact with the ground at any other point than at the bump. Hence the reaction of the ground upon the roller acts at the bump alone.[8] Force equilibrium gives

$$N_y - W = 0 \qquad \text{and} \qquad N_x + F = 0$$

Moment equilibrium will give us a third equation sufficient to determine the three unknowns N_x, N_y and, the really important one, F. I take moments about the center of the roller and immediately observe that the resultant of N_x and N_y must pass through the center of the roller since only it produces a moment about our chosen reference point. That is, the orientation of the reaction force shown below on the left violates moment equilibrium. The reaction force must be directed as shown in the middle figure.

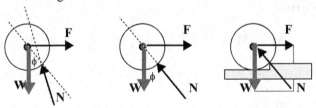

---

8. It is impossible give you a cookbook rule on how to read phrases like *just start*. Effective working knowledge comes with the exercising of the language. Furthermore I can give you no assurance that an author of another text, will use the exact same phrase to indicate this condition.

Furthermore, we require the three forces to sum to zero to satisfy force equilibrium, so: $\mathbf{F} + \mathbf{W} + \mathbf{N} = \mathbf{0}$.

This is shown graphically in the figure at the far right above. From this we obtain what we were asked to show. (A bit of analytical geometry leads to both relationships). Now from this we can state another rule, namely

- **For an isolated system subject to but three forces, a** *three-force system,* **the three forces must be concurrent. That is, their lines of action must all run together and intersect at a common point.**

Note the priority of the equilibrium requirements in fixing a solution. We might have begun worrying about the frictional force, or something akin to a frictional force, acting at the bump, resisting motion. But our analysis says there is *no* frictional force! The reaction is perpendicular to the plane of contact; the latter is tangent to the surface at the bump and perpendicular to the radius. How can this be? Nothing was said about "assume friction can be neglected" or "this urchin is pulling the roller up over a bump in the ice on the pond" No, because nothing need be said! Moment equilibrium insists that the reaction force be directed as shown. Moment equilibrium has top priority. The Platonists win again; it's mostly, if not all, in your mind.

Resultant Force and Moment

For static equilibrium of an isolated body, the resultant force and the resultant moment acting on the body must vanish. These are vector sums.

Often it is useful to speak of the resultant force and moment of some **subset** of forces (and moments). For example, a moment vector can be spoken of as the resultant of its, at most, three scalar components. In the figure, the

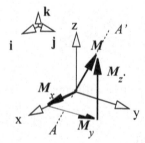

moment or torque about the inclined axis *AA'* is the resultant or vector sum of the three vectors $\mathbf{M}_x$, $\mathbf{M}_y$ and $\mathbf{M}_z$ each of which can be written as the product of a scalar magnitude and a unit vector directed along the appropriate coordinate axis. That is, we can write.

$$\mathbf{M} = \mathbf{M}_x + \mathbf{M}_y + \mathbf{M}_z = M_x \cdot \mathbf{i} + M_y \cdot \mathbf{j} + M_z \cdot \mathbf{k}$$

A moment in itself can be spoken of as the resultant of a force but, while this phrasing is formally correct, it is rarely used. Instead we speak of the *moment at a point due to a force* or the *moment about a point of the force....* as in "the moment, about the point O of the force $\mathbf{F}_A$ is given by the product of (1) the perpendicular distance from the point O to the line of action of the force and (2) the magnitude of the force. Its direction is given by the *right-hand rule.*

The *right-hand rule* is one of those oddities in science and engineering. It is often stated as

- **The direction of the moment of a force about a point is the same as the direction of advance of a right handed screw when the screw is oriented perpendicular to the plane defined by the line of action of the force and the reference point for the moment.**

This is a mouthful but it works. It meets the need to associate a direction with the turning effect of a force.

We can seemingly avoid this kind of talk by defining the moment as the vector cross product of a position vector from the reference point to *any point* on the line of action of the force. But we are just passing the buck; we still must resort to this same way of speaking in order to define the direction of the vector cross product.

It is worth going through the general definition of moment as a vector cross product[9]. Some useful techniques for calculating moments that avoid the need to find a perpendicular distance become evident

The magnitude of the vector cross product, that is the magnitude of the moment of the force about the point O above is

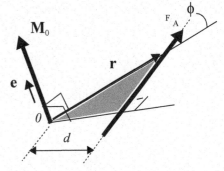

$$|\mathbf{r} \times \mathbf{F}_A| = |\mathbf{M}_O| = |\mathbf{r}| \cdot |\mathbf{F}_A| \cdot sin\phi$$

or, striking the boldface to indicate scalar magnitude alone $M_O = r \cdot F_A \cdot sin\phi$

This is in essence the definition of the vector cross product. The direction of the moment is indicated on the figure by the unit vector **e**. Note: e is commonly used to represent a unit vector. Its heritage is German; *eine* is one.

Note that we recover the more specialized definition of the magnitude of the moment, that which speaks of perpendicular distance from the reference point to the line of action of the force, by writing

$$d = r \cdot sin\phi \qquad \text{so that} \qquad M_O = d \cdot F_A$$

Observe that I could have interpreted the magnitude of the moment as the product of r and the component of the force perpendicular to the position vector, namely F$sin\phi$.

---

9. *Vector* and *cross* are redundant when used to describe "product". We ought to speak only of *the vector product* or *the cross product*. Similarly, the *dot product* of two vectors can be called *the scalar product*. It would be again redundant to speak of *the scalar, dot product*. We speak redundantly here in order to emphasize that the outcome of a cross product is a vector.

In evaluating the cross product, you must take care to define the angle ϕ as the included angle between the position vector and the force vector when the two are placed "tail-to-tail". ϕ is the angle swept out when you swing the position vector around to align with the force vector, moving according to the right hand rule.

The payoff of using the cross product to evaluate the moment of a force with respect to a point is that you can choose a position vector from the point to *any point on the line of action* of the force. In particular, if you have available the scalar components of some position vector and the scalar components of the force, the calculation of both the magnitude and direction of the moment is a machine-like operation.

Exercise 2.9

Show that the moment about point A due to the tension in the cable DB is given by

$$\mathbf{M} = (-0.456 \cdot \mathbf{i} + 0.570 \cdot \mathbf{k}) \cdot L \cdot F_D$$

where *i, j, k* are, as usual, three unit vectors directed along the three axes x, y, z.

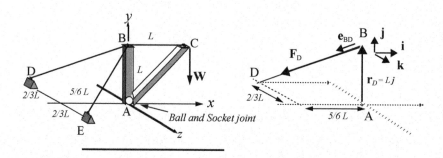

The approach is reductive and mechanical. I will use the vector machinery of the cross product, first expressing a position vector and the force in terms of their x,y,z components using the scalar magnitude/unit vector mode of representation for each. For the position vector I write, enjoying the freedom I have to select any point on the line of action of the force as the head of the vector:

$$\mathbf{r}_D = L \cdot \mathbf{j}$$

Simple enough! But now matters become more complex. I write the tension in the cable as the product of its magnitude, F_D and a unit vector directed along its line of action, along the line running from B to D. Constructing the unit vector requires the application of the pythagorean theorem and other bits of analytical geometry. From the Pythagorean Theorem, the length of the cable BD is

$$L_{BD} = [(5/6)^2 + (1)^2 + (2/3)^2]^{1/2} \cdot L$$

The components of the unit vector, e_{BD}, from B to D are then $-(5/6)L/L_{BD}$ along x, $-1L/L_{BD}$ along y, and $-(2/3)L/L_{BD}$ along z. So:

$$e_{BD} = -(5/6)(L/L_{BD})\mathbf{i} - 1(L/L_{BD}) \cdot \mathbf{j} - (2/3)(L/L_{BD}) \cdot \mathbf{k}$$

or, with L_{BD} defined above and calculating to three significant figures[10],

$$e_{BD} = -0.570 \cdot \mathbf{i} - 0.684 \cdot \mathbf{j} - 0.456 \cdot \mathbf{k}$$

The moment of the tension in the cable BD about A is then obtained from the cross product:

$$\mathbf{M} = \mathbf{r} \times \mathbf{F}_D = L \cdot \mathbf{j} \times (-0.570 \cdot \mathbf{i} - 0.648 \cdot \mathbf{j} - 0.456 \cdot \mathbf{k}) \cdot F_D$$

Our cross product is now a collection of three cross products. The machinery for doing these includes

$$\mathbf{i} \times \mathbf{j} = \mathbf{k} \qquad \mathbf{j} \times \mathbf{k} = \mathbf{i} \qquad \mathbf{k} \times \mathbf{i} = \mathbf{j}$$

and recognition that the cross product is sensitive to the order of the two vectors, for example $j \times i = -k$, and that the vector product of a unit vector with itself is zero. Cranking through, we obtain

$$\mathbf{M} = (-0.456 \cdot \mathbf{i} + 0.570 \cdot \mathbf{k}) \cdot L \cdot F_D$$

- Observe that the moment has a component along the x axis (negatively directed), another along the z axis but not along the y axis; the force produces no turning effect about the vertical axis. Indeed, the force intersects the y axis.

---

10. Three significant figures are generally sufficient for most engineering work. You are responsible for rounding off the 8, 16 or 20 numbers that contemporary spread sheets and other calculating machinery generate. You must learn how to bite off what you can't possibly chew.

- Observe that you can deduce from this result the perpendicular distance from *A* to the line of action of the force by recasting the vector term as a unit vector. The factor introduced to accomplish this when associated with the length *L* then defines the perpendicular distance. That is, we write, noting that $[.456^2 + .570^2]^{1/2} = .730$

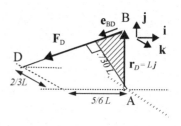

$$\mathbf{M} = (-0.627 \cdot \mathbf{i} + 0.781 \cdot \mathbf{k})(0.730 \cdot L) \cdot F_D$$

and so the perpendicular distance from *A* to the line of action must be 0.730L.

- Finally, observe that I could have constructed a position vector from *A* to *D*, namely

$$\mathbf{r} = -(5/6) \cdot L \cdot \mathbf{i} - (2/3) \cdot L \cdot \mathbf{k}$$

and used this to compute the moment. (See problem 2.8)

Couple

There exists one particularly important resultant moment, that due to two equal and opposite forces – so particular and important that we give it its own name – *couple*. To see why this moment merits singling out, consider the resultant moment about the point O of the two equal and opposite forces F_A and F_B.

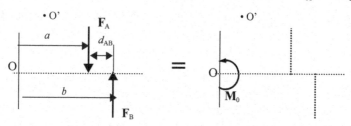

We can write the scalar magnitude of the moment, taking counter clockwise as positive,

$$M_O = -a \cdot F_A + b \cdot F_B = (b - a) \cdot F_A$$

where we have explicitly made use of the fact that the forces are equal and opposite by taking $F_B = -F_A$. Note that this equation is written only in terms of the scalar magnitudes of the forces and the moment. The figures on the left and right show two *equivalent systems of forces*.

Note that $(b-a)$ is just the (perpendicular) distance, d_{AB} between the two parallel lines of action of the two forces, so if I were to ask "what is the moment of the two forces about some other point?" say O', my response would be the same. In fact

- **the moment due to two equal and opposite forces, *a couple* is invariant with respect to choice of reference point; its value is given by the product of the perpendicular distance between the two lines of action of the two forces and the value of the force.** The resultant force due to the two equal and opposite forces is zero.

Exercise 2.10

Show that, for equilibrium of the cantilever beam loaded with two equal and oppositely directed forces as shown, the reaction at the wall is a couple of magnitude Fd and having a direction "out of the paper" or counterclockwise.

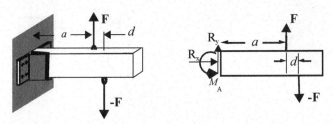

We isolate the cantilever beam, showing all the applied and reaction forces (and couples) acting. This is shown at the right. Note the symbol, the circular arrow. It represents the unknown couple or moment *acting at the point A*, assumed here to be acting counterclockwise. Force equilibrium requires

$$R_x = 0; \qquad\qquad R_y + F - F = 0$$

These immediately lead to the conclusion that the reaction force at A is zero. R_y as well as R_x are zero.

Moment equilibrium *about the point A*[11] requires

$$M_A - F \cdot d = 0 \qquad \text{so} \qquad M_A = F \cdot d$$

We say *The reaction at A is the* **couple Fd.**

---

11. Note the difference between the two phrases *moment acting at the point or about the point* and *moment **equilibrium** about the point*. We say there is a moment (couple) acting *at* the point and then, in the next breath, say that the resultant moment *about* the point must vanish for equilibrium to be satisfied. There is nothing inconsistent here. It is essential that you understand the difference in these two expressions.

The exercise above might strike you as overly abstract and useless, almost a tautology, as if we proved nothing of worth. Don't be fooled. The purpose of the exercise is not algebraic analysis but rather to move you to accept the existence of *couple* as an entity in itself, a thing that is as real as *force*. You may not sense a couple the way you feel a force but in Engineering Mechanics the former is just as lively and substantial a concept as the latter.

Note too how if the beam had been clamped to a wall at its other end, at the right rather than at the left, the reaction at the wall would still be a *couple* of magnitude *Fd* directed counterclockwise.

Exercise 2.11

What if the beam of the last exercise is supported at the wall by two frictionless pins A,B? What can you say about the reaction forces at the pins?

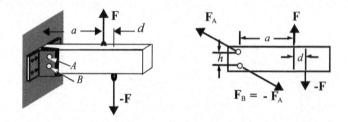

We can say that the two forces acting on the pins, say F_A, F_B, are equal and opposite. We can say that they are equivalent to a *couple* of magnitude Fd and of direction counterclockwise. All of this follows from both force and moment equilibrium requirements applied to the beam when isolated from the wall to which it is pinned. We can say nothing more.

Well, that's not quite the case: While we cannot say what their magnitude is, we can say that they are *at least* as big as the product $F(d/h)$. That is $F_A = F_B \geq F \cdot (d/h)$ because *h* is the maximum possible perpendicular distance that can be drawn between any two parallel lines of action drawn through *A* and *B*.

The couple *Fd* in these few exercises is an *equivalent system* – equivalent to that of two equal and opposite forces separated a distance *d*. In fact, the notion of *equivalent force system* is closely wrapped up with the requirements of force and of moment equilibrium. We can replace one equivalent force system by another in our equilibrium deliberations and the results of our analysis will be the same because

- **equivalent systems have the same resultant force and the same resultant moment about any, arbitrarily chosen point.**

Furthermore, now that we allow a couple vector to be a thing in itself:

- every system of any number of force vectors and couple vectors can be represented by an equivalent system consisting of a *single* force vector together with a *single* couple vector.

For example, the three systems shown below are all equivalent

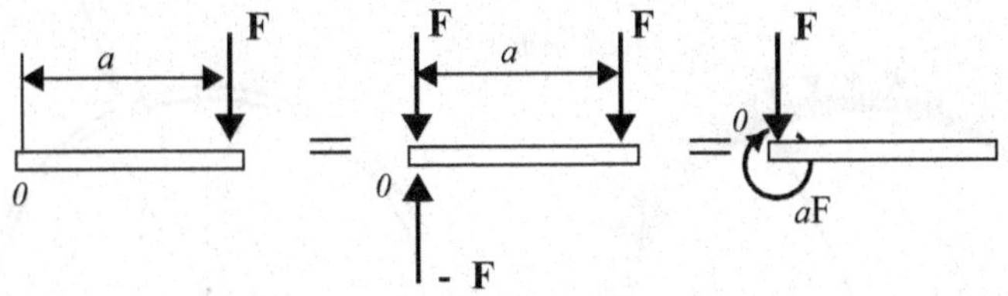

Exercise 2.12

Show that for the load W, uniformly distributed over the span L, (a) an equivalent system is a force vector of magnitude W directed downward acting at the center of the span and no couple; (b) another is a force vector of magnitude W directed downward acting at the left end of the span and a couple of magnitude W(L/2) directed clockwise; another is.....

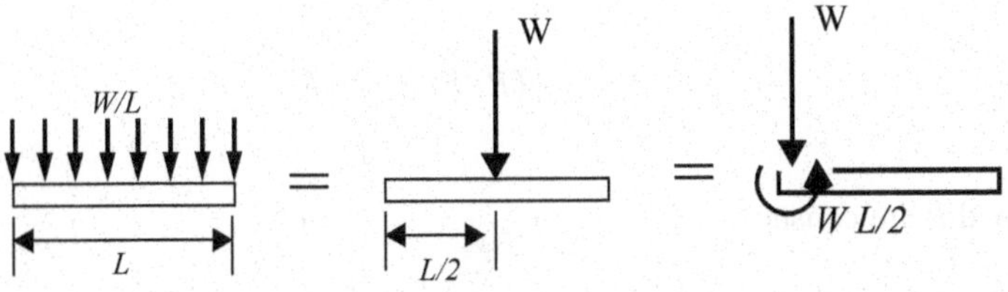

For parts (a) and (b) the resultant force is clearly the total load *W*. In fact, regardless of how the load is distributed over the span, the resultant, equivalent force, will have magnitude *W* and be directed vertically downward. The location of its line of action is a matter of choice. Different choices, however imply different moments or couples — the other ingredient of our equivalent system.

Taking midspan as our reference point, (a), we see that the load, if uniformly distributed, on the right will produce a moment clockwise about center span, that on the left will produce a moment *counterclockwise* about center span. The magnitudes of the two moments will be equal, hence the resultant moment of the uniformly distributed load with reference to center span is zero.

Taking the left end of the span as our reference point, (b), we can make use of the equivalence displayed in the preceding figure on the preceding page where now *a* is read as the distance *L/2* between the vertical load *W* acting downward at center span and the left end of the span.

Exercise 2.13

Show that a radially directed force per unit length, α, with units newtons/meter, uniformly distributed around the circumference of the half circle is equivalent to a single force vector of magnitude 2α R, where R is the radius of the half circle, and no couple.

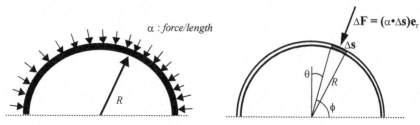

α : *force/length*

$$\Delta \mathbf{F} = (\alpha \cdot \Delta \mathbf{s}) \mathbf{e}_r$$

We integrate. We sum up vertical components of differential elements of the distributed force around the circumference.

A differential element of force, ΔF, acting on the differential length of circumference Δs is where the unit vector $\mathbf{e}_r$ is taken positive inward. The vertical $$\Delta \mathbf{F} = (\alpha \cdot \Delta s) \cdot \mathbf{e}_r$$

component of this is given by $\alpha \Delta s \cos\theta$ which, noting that $\theta = (\pi/2) - \phi$, can be written $\alpha \sin\phi \Delta s$. Now Δs is just $R\Delta\phi$ so we can write

$$F_{down} = \int_0^\pi \alpha \sin\phi (R d\phi) = 2\alpha R$$

where the α and R are constants.

Design Exercise 2.1

The exercise set out below and the others like it to follow are not complete. It is a design exercise and, as such, differs from the problems you have been assigned up to now. It, and the other design exercises to follow, is different in that there is no single right answer. Although these exercises are keyed to specific single-answer problems in the text and are made to emphasize the fundamental concepts and principles of the subject, they are open-ended. The responses you construct will depend upon how you, your classmates, and your recitation instructor flesh out the task.

In effect, we want you to take responsibility in part for defining the problem, for deciding which constraints and specifications are critical, and setting the context for evaluating possible "solutions". Design is the essence of engineering and the act of design includes formulating problems as much as solving them, negotiating constraints as well as making sure your solutions respect them, and teamwork as well as individual competence.

Hospital-bed Wheel Size

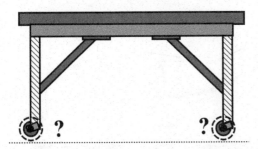 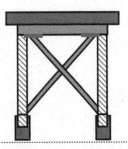

Your task is to do a first-cut analysis in support of the design of a new, lightweight, mobile hospital bed. You know that the bed will be used to transport patients indoors on caster type wheels over relatively smooth terrain but there will be some small obstacles and bumps it must traverse without discomfort to the patient. A single attendant should be able to push the bed to its destination. Develop a rationale for fixing the size of the wheels and use it to determine a range of possible diameters.

Design Exercise 2.2

Your boss wants to diversify. The market for portable stanchions for volleyball nets has diminished over the years with the introduction of more and more cable television stations and the opening of the information highway. People are spending most of their time just lying around, scanning channels. He figures there will be a growing demand for hammocks and the stanchions to support them.

The stanchions are to be portable. He wants them to be able to support the hammock's occupant (occupants?) without fastening the stanchion to the ground. He envisions the ground to be a level surface with most users indoors. The area, the *footprint*, of the hammock and stanchions is limited and you are to assume that there are no walls to tie any supports to.

Your job is to identify the most important design parameters to ensure that the free standing stanchions hold the hammock in the desired form - indicated below. In this you want to construct estimates of the required weight of the stanchions, coefficients of friction, and explore limits on, and ranges of, pertinent dimensions

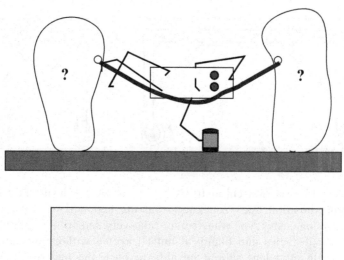

Top vue -- System footprint

of the hammock, the stanchions, as well as the height of attachment and distance between them.

2.3   Problems - Static Equilibrium

2.1   *Estimate* the weight of a compact automobile; the weight of a fully loaded trailer truck.

2.2   *Estimate* the weight of a paper clip; the total weight borne by a book shelf one meter long, fully packed with books; the weight of the earth's moon.

2.3   *Estimate* the angle of a hill upon which you can safely park your car under dry road conditions; under icy conditions.

2.4   *Estimate* the maximum shearing force you can apply to the horn of some domesticated animal by means of **Leavitt's V Shape Blade Dehorning Clipper**.

LEAVITT'S V SHAPE BLADE DEHORNING CLIPPER

FOR OTHER VETERI-
NARY INSTRUMENTS,
SEE DRUG DEPART-
MENT.

No. 9R4640   This illustration represents our V shape blade Dehorning Clipper, which cuts all around the horn as handles are closed. Knives cannot interlock, or cut into each other. Has double power and is guaranteed to be superior to any other dehorner made. This style dehorner is made in the large size (No. 3) only, but will clip any size horn from cattle of any age, smooth and clean. In opening the blades, the handles do not go far enough apart to prevent the operator having ample purchase, and twice the power of any other dehorner made. This is a very desirable feature, as it has power enough to clip any large horn, with perfect ease to the operator.

ANOTHER IMPORTANT FEATURE.

It will be seen that in closing the clipper, the same power (the cogged handles) which drives the cogged plunger down on sliding blade, thus making a machine with two movable blades, with power from both sides of cogged handles. This clipper is made with only three bolts. One small bolt fastens cogged plunger to sliding knife, and acts as a stop for blade, making a 4-inch opening in large machine, which is large enough to admit any horn. The other two bolts fasten end blade.

WE CHALLENGE THE WORLD TO PRODUCE A CLIPPER WITH MERITS THAT EQUAL THE LEAVITT LATEST IMPROVED V SHAPE BLADE DEHORNING CLIPPER.

2.5   *Show that* if the coefficient of friction between the block and the plane is 0.25, the force required to just start the block moving **up** the 40° incline is $F = 1.38\ W$ while the force required to hold the block from sliding **down** the plane is $F = 0.487\ W$.

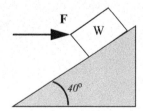

2.6   A block slides down an plane inclined at 20°. *What if* the coefficient of friction is doubled; at what angle of incline will the block begin to slide down the plane?

2.7   *Estimate* the difference in water pressure between the fourth and first floors of bldg. 1.

2.8   *Show that* if, in exercise 2.8, you take a position vector from A to D in computing the moment due to the tension in the cable BD, rather than from A to B as was done in the text, you obtain the same result for the moment of the tension with respect to the point A.

2.9   What do you *need to know* to estimate the torque a driver of a compact automobile must exert to turn the steering wheel (no power steering) while the vehicle is at rest?

2.10   *Estimate* the "live" floor loading in lecture 10-270 during the first 8.01 class of the semester; during the last class of the semester.

2.11   *What if* the urchin of exercise 2.7 pulling the roller grows up and is able to pull on the handle at an angle of θ up from the horizontal; Develop a nondimensional expression for the force she must exert to just start the roller over the curb.

2.12   *Show that*, for the block kept from sliding down the inclined plane by friction alone, that the line of action of the resultant of the normal force (per unit area) distributed over the base of the block must pass through the intersection of the line of action of the weight vector and the line of action of the friction force. What happens when the line of action of the weight falls to the right of point B?

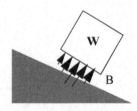

2.13   The rigid beam carries a load P at its right end and is supported at the left end by two (frictionless pins). The pin at the top is pulled upwards and held in place by a cable inclined at a 45 degree angle with the horizontal.

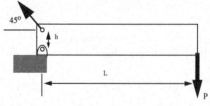

Draw a free body diagram of the beam, isolated from its environment i.e., show all the forces acting on the beam alone; show all relevant dimensions; show a reference cartesian axes system.

2.14 The reaction force at B, of the wall upon the ladder, is greater than, equal to, or less than the weight, W, of the ladder alone?

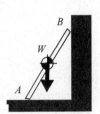

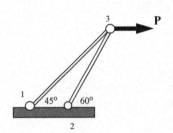

2.15 Isolate pin 3, showing the forces acting on the pin due to the tension (or compression) in the two members.

Find the forces in the members in terms of **P**.

2.15 The truss structure shown carries a load **P** and is supported by a cable, *BC*, and pinned at *D* to the wall. Determine the force in the cable *BC* and the reaction force at *D*.

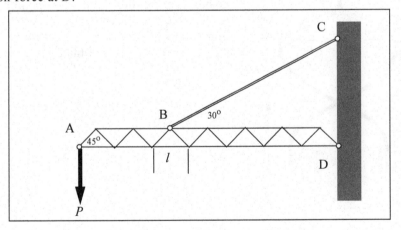

2.16   Isolate pin 4, showing the forces acting on the pin due to the tensions (or compressions) in the three members.

1) Write out consequences of the requirements for static equilibrium of pin 4.

2) Can you solve for the member forces in terms of **P** (and the given angles)?

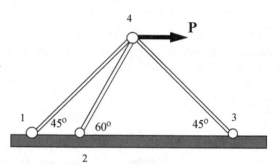

2.17   Estimate the weight of a fully loaded MBTA bus traveling down Mass. Avenue. Estimate an upper bound on the loading per unit length of span of the Harvard Bridge if clogged with buses, both ways.

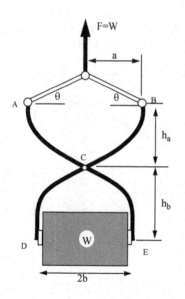

2.18 The vice grip shown is made of two formed steel members pinned at C. Construct an expression for the force component which compresses the block at the bottom in terms of the weight of the block W and the dimensions shown. Express your result in terms of nondimensional factors.

If $\theta = 30°$ $a = b = h_a = h_b$, What must the coefficient of friction be to ensure the block does not slide out of the grip?

3

Internal Forces and Moments

3.1   Internal Forces in Members of a Truss Structure

We are ready to start talking business, to buy a loaf of bread. Up until now we have focused on the rudimentary basics of the language; the vocabulary of force, moment, couple and the syntax of static equilibrium of an isolated particle or extended body. This has been an abstract discourse for the most part. We want now to start speaking about "extended bodies" as *structural members*, as the building blocks of *truss structures, frame structures, shafts and columns* and the like. We want to go beyond questions about forces and moments required to satisfy equilibrium and ask "...When will this structure break? Will it carry the prescribed loading?"

We will discover that, with our current language skills, we can only answer questions of this sort for one type of structure, the *truss structure*, and then only for a subset of all possible truss structures. To go further we will need to broaden our scope, beyond the requirements of force and moment equilibrium, and analyze the *deformations and displacements* of extended bodies in order to respond to questions about load carrying ability regardless of the type and complexity of the structure at hand - the subject of a subsequent chapter. Here we will go as far as we can go with the vocabulary and rules of syntax at our disposal. After all, the requirements of force and moment equilibrium still must be satisfied whatever structure we confront.

A *truss structure* is designed, fabricated, and assembled such that its members carry the loads in tension or compression. More abstractly, **a truss structure is made up of straight, two-force members, fastened together by frictionless pins; all loads are applied at the joints.**

Now, we all know that there is no such thing as a truly frictionless pin; you will not find them in a suppliers catalogue. And to require that the loads be applied at the joints alone seems a severe restriction. How can we ensure that this constraint is abided by in use?

We can't and, indeed, frictionless pins do not exist. This is not to say that there are not some ways of fastening members together that act more like frictionless pins than other ways.

What *does* exist inside a truss structure are forces and moments of a quite general nature but the forces of tension and compression within the straight members are the most important of all if the structure is designed, fabricated, and assembled according to accepted practice. That is, the loads within the members of a truss structure may be approximated by those obtained from an analysis of an abstract representation (as straight, frictionless pinned members, loaded at the joints alone) of the structure. Indeed, this abstract representation is what serves as the basis for the design of the truss structure in the first place.

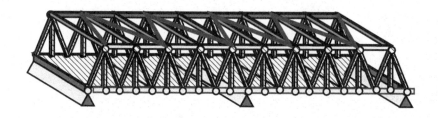

Any member of the structure shown above will, in our abstract mode of imagining, be in either tension or compression – a state of *uniaxial loading*. Think of having a pair of special eyeglasses – truss seeing glasses – that, when worn, enable you to see all members as straight lines joined by frictionless pins and external forces applied at the joints as vectors. This is how we will usually sketch the truss structure, as you would see it through such magical glasses.

Now if you look closer and imagine cutting away one of the members, say with circular cross-section, you would see something like what's shown: This particular member carries its load, F, in tension; The member is being stretched.

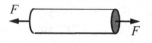

If we continue our imagining, increasing the applied, external loads slowly, the tension in this member will increase proportionally. Eventually, the member will fail. Often a structure fails at its joints. We rule out this possibility here, assuming that our joints have been *over-designed.* The *way* in which the member fails, as well as the tensile force at which it fails, depends upon two things: The cross-sectional area of the member and the material out of which it is made.

If it is grey cast iron with a cross sectional area of 1.0 in^2, the member will *fracture*, break in two, when the tensile force approaches 25,000 *lb*. If it is made of aluminum alloy 2024-T4 and its area is $600mm^2$ it will *yield*, begin to deform *plastically*, when the tensile forces approaches 195,000 *Newtons*. In either case there is some magnitude of the tensile force we do not want to exceed if we wish to avoid failure.

Continue on with the thought experiment: Imagine that we replaced this member in our truss structure with another of the same length but twice the cross-sectional area. What tensile load can the member now carry before failure?

In this imaginary world, the tensile force required for failure will be twice what it was before. In other words, we take the measure of failure for a truss member in tension to be the *tensile (or compressive) stress* in the member where the stress is defined as the magnitude of the force divided by the cross-sectional area of the member. In this we assume that the force is *uniformly distributed* over the cross-sectional area as shown below.

$$\sigma = F/A$$

Further on we will take a closer look at how materials fail due to internal forces, not just tensile and compressive. For now we take it as an empirical observation and operational heuristic that to avoid fracture or yielding of a truss member we want to keep the **tensile or compressive stress** in the member below a certain value, a value which depends primarily upon the material out of which the member is made. (We will explore later on when we are justified taking a failure stress in uniaxial compression equal to the failure stress in uniaxial tension.) It will also depend upon what conventional practice has fixed for a *factor of safety*. Symbolically, we want

$$F/A < (F/A)_{failure} \qquad \text{or} \qquad \sigma < \sigma_{failure}$$

where I have introduced the symbol σ to designate the uniformly distributed stress.

Exercise 3.1

If the members of the truss structure of Exercise 2.7 are made of 2024-T4 Aluminum, hollow tubes of diameter 20.0 mm and wall thickness 2.0 mm, estimate the maximum load P you can apply before the structure yields.[1] In this take θ_A, θ_B to be 30^0, 60^0 respectively

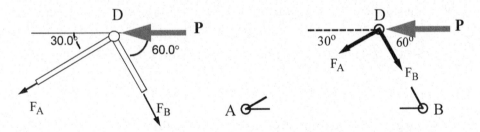

Rather than picking up where we left off in our analysis of Exercise 2.7., we make an alternate isolation, this time of joint, or *node D* showing the unknown member forces directed along the member. By convention, we assume that both members are in tension. If the value for a member force comes out to be negative,

---

1. Another failure mode, other than yielding, is possible: Member *AD* might *buckle*. We will attend to this possibility in the last chapter.

we conclude that the member is in compression rather than tension. This is an example of a convention often, but not always, adopted in the analysis and design of truss structures. You are free to violate this norm or set up your own, but beware: It is your responsibility to note the difference between your method and what we will take as conventional and understood without specification.

Force equilibrium of this *node as particle* then provides two scalar equations for the two scalar unknown member forces. We have

$$-F_A \cdot \cos 30^o + F_B \cdot \cos 60^o - P = 0 \qquad -F_A \cdot \sin 30^o - F_B \cdot \sin 60^o = 0$$

From these, we find that member *AD* is in *compression*, carrying a load of $(\sqrt{3}/2)P$ and member *BD* is in *tension*, carrying a load $P/2$. The stress in each member is the force divided by the cross-sectional area where I have approximated the area

$$A = \pi \cdot (20 \times 10^{-3}) \cdot (2 \times 10^{-3}) m^2 = 40\pi \times 10^{-6} m^2$$

of the cross-section of the thin-walled tube as a rectangle whose length is equal to the circumference of the tube and width equal to the wall thickness.

Now the compressive stress in member *AD* is greater in magnitude than the tensile stress in member *BD* – about 1.7 times greater – thus member *AD* will yield first. This defines the mode of failure. The *compressive stress* in *AD* is

$$\sigma = (\sqrt{3}/2) \cdot P/A$$

which we will say becomes excessive if it approaches 80% of the value of the stress at which 2024-T4 Aluminum begins to yield in uniaxial tension. The latter is listed as 325 *MegaPascals* in the handbooks[2].

We estimate then, that the structure will fail, due to yielding of member *AD*, when

$$P = (2/\sqrt{3}) \cdot (0.80) \cdot (40\pi \times 10^{-6} m^2) \cdot (325 \times 10^6 N/m^2) = 37,700N$$

***

I am going to now alter this structure by adding a third member *CD*. We might expect that this would *pick up some of the load*, enabling the application of a load *P* greater than that found above before the *onset of yielding*. We will discover that we cannot make this argument using our current language skills. We will find that we need new vocabulary and rules of syntax in order to do so. Let us see why.

---

2. A Pascal is one Newton per Square meter. Mega is 10^6. Note well how the dimensions of stress are the same as those of pressure, namely, force per unit area. See Chapter 7 for a crude table of failure stress values.

Exercise 3.2

*Show that if I add a third member to the structure of **Exercise 3.1** connecting node D to ground at C, the equations of static equilibrium do not suffice to define the tensile or compressive forces in the three members.*

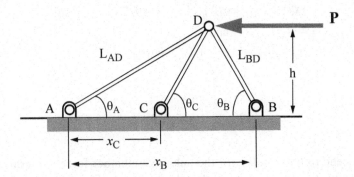

I isolate the system, starting as I did when I first encountered this structure, cutting out the whole structure from its supporting pins at *A*, *B* and *C*. The free

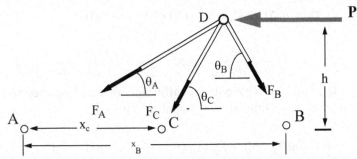

body diagram above shows the direction of the unknown member forces as along the members, a characteristic of this and every truss structure. Force equilibrium in the horizontal direction and vertical direction produces two scalar equations:

$$-F_A \cdot \cos\theta_A - F_C \cdot \cos\theta_C + F_B \cdot \cos\theta_B - P = 0$$

and

$$F_A \cdot \sin\theta_A + F_B \cdot \sin\theta_B + F_C \cdot \sin\theta_C = 0$$

At this point I note that the above can be read as two equations in three unknowns — the three forces in the members — presuming we are given the angles θ_A, θ_B, and θ_C, together with the applied load *P*. We clearly need another equation.

Summing moments about A I can write

$$x_B \cdot F_B \cdot sin\theta_B + x_C \cdot F_C \cdot sin\theta_C - h \cdot P = 0$$

$$x_C = L_{AD} \cdot cos\theta_A - L_{CD} \cdot cos\theta_C$$

where   and

$$x_B = L_{AD} \cdot cos\theta_A + L_{BD} \cdot cos\theta_B$$

and

$$h = L_{AD} \cdot sin\theta_A = L_{CD} \cdot sin\theta_C = L_{BD} \cdot sin\theta_B$$

I now proceed to try to solve for the unknown member forces in terms of some or all of these presumed given geometrical parameters. First I write the distances

$$x_C = h \cdot [(cos\theta_A / sin\theta_A) - (cos\theta_C / sin\theta_C)]$$

x_C and x_B as   and

$$x_B = h \cdot [(cos\theta_A / sin\theta_A) + (cos\theta_B / sin\theta_B)]$$

or

$$x_C = h \cdot \left[\frac{sin(\theta_C - \theta_A)}{sin\theta_A \cdot sin\theta_C}\right] \quad \text{and} \quad x_B = h \cdot \left[\frac{sin(\theta_A + \theta_B)}{sin\theta_A \cdot sin\theta_B}\right]$$

With these, the equation for moment equilibrium about A becomes after substituting for the x's and canceling out the common factor h,

$$F_B \cdot sin(\theta_A + \theta_B) + F_C \cdot sin(\theta_C - \theta_A) - P \cdot sin\theta_A = 0$$

It appears, at first glance that we are in good shape, that we have three scalar equations – two from force equilibrium, this last from moment equilibrium – available to determine the three member forces, F_A, F_B and F_C. We proceed by eliminating F_A, one of our unknowns from the equations of force equilibrium. We will then be left with two equations – those we derive from force and moment equilibrium – for determining F_B and F_C.

We multiply the equation expressing force equilibrium in the horizontal direction by the factor $sin\theta_A$, that expressing force equilibrium in the vertical direction by the factor $cos\theta_A$ then add the two equations and obtain

$$F_B \cdot (sin\theta_A \cdot cos\theta_B + cos\theta_A \cdot sin\theta_B) + F_C \cdot (sin\theta_C \cdot cos\theta_A - sin\theta_A \cdot cos\theta_C) - P \cdot sin\theta_A = 0$$

This can be written, using the appropriate trig identities,

$$F_B \cdot sin(\theta_A + \theta_B) + F_C \cdot sin(\theta_C - \theta_A) - P \cdot sin\theta_A = 0$$

which is identical to the equation we obtained from operations on the equation of moment equilibrium about A above. This means we are up a creek. The equation of moment equilibrium gives us no new information we did not already have from the equations of force equilibrium. We say that the equation of moment equilibrium is linearly dependent upon the latter two equations. We cannot find a unique solution for the member forces. We say that **this system of equations is linearly dependent**. We say that **the problem is statically, or equilibrium, indeterminate.** The equations of equilibrium do *not* suffice to enable us to find a unique solution for the unknowns. Once again, the meaning of the word indeterminate is best illustrated by the fact that we can find many, many solutions for the member forces that satisfy equilibrium.

This time there are no special tricks, no special effects hidden in subsystems, that would enable us to go further. That's it. We can not solve the problem. Rather, we have solved the problem in that we have shown that the equations of equilibrium are insufficient to the task.

Observe

- That the forces in the members might depend upon how well a machinist has fabricated the additional member *CD*. Say he or she made it too short. Then, in order to assemble the structure, you are going to have to pull the node *D* down toward point *C* in order to fasten the new member to the others at *D* and to the ground at *C*. This will mean that the members will experience some tension or compression even when the applied load is zero[3]! We say the structure is *preloaded*. The magnitudes of the preloads will depend upon the extent of the incompatibility of the length of the additional member with the distance between point *C* and *D*

- We don't need the third member if the load *P* never comes close to the failure load determined in the previous exercise. The third member is *redundant*. In fact, we could remove any *one* of the other two members and the remaining two would be able to support a load *P* of some significant magnitude. With three members we have a *redundant structure*. **A redundant structure is most often synonymous with a statically indeterminate system of equations.**

- I could have isolated joint *D* at the outset and immediately have recognized that only two linearly independent equations of equilibrium are available. Moment equilibrium would be *identically satisfied* since all force vectors intersect at a common point, at the node *D*.

---

3. This is one reason why no engineering drawing of structural members is complete without the specification of tolerances.

In the so-called "real world", some truss structures are designed as redundant structures, some not. Why you might want one or the other is an interesting question. More about this later.

Statically determinate trusses can be quite complex, fully three-dimensional structures. They are important in their own right and we have all that we need to determine their member forces— namely, the requirements of static equilibrium.

Exercise 3.3

Construct a procedure for calculating the forces in all the members of the statically determinate truss shown below. In this take α = √3

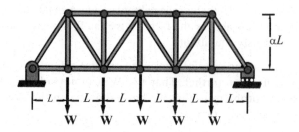

1. *We begin with an isolation of the entire structure:*

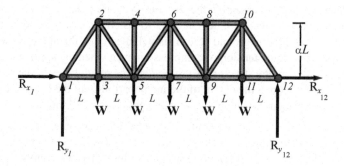

2. *Then we determine the reactions at the supports.*

This is not always a necessity, as it is here, but generally it is good practice. Note all of the strange little circles and shadings at the support points at the left and right ends of the structure. The icon at the left end of the truss is to be read as meaning that:

- the joint is frictionless and
- the joint is restrained in both the horizontal and vertical direction, in fact, the joint can't move in any direction.

The icon at the right shows a frictionless pin at the joint but it itself is sitting on more frictionless pins. The latter indicate that the joint is free to move in the horizontal direction. This, in turn, means that the horizontal component of the reaction force at this joint, Rx_{12} is zero, a fact crucial to the *determinancy* of the problem. The shading below the row of circles indicates that the joint is *not* free to move in the vertical direction.

From the symmetry of the applied loads, the total load of 5W is shared equally at the supports. Hence, the vertical components of the two reaction forces are

$$Ry_1 = Ry_{12} = 5\,W/2.$$

Both of the horizontal components of the reaction forces at the two supports must be zero if one of them is zero. This follows from the requirement of force equilibrium applied to our isolation.

$$Rx_1 = Rx_{12} = 0.$$

3. *Isolate a joint at which but* **two** *member forces have yet to be determined and apply the equilibrium requirements to determine their values.*

There are but two joints, the two support joints that qualify for consideration this first pass through the procedure. I choose to isolate the joint at the left support. Equilibrium of force of node # 1 in the horizontal and vertical direction yields the two scalar equations for the two unknown forces in members 1-2 and 1-3. In this we again assume the members are in tension. A negative result will then indicate the member is in compression. The proper way to speak of this feature of our isolation is to note how "the members in tension pull on the joint".

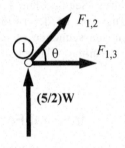

Equilibrium in the x direction and in the y direction then requires:

$$F_{1,2} \cdot \cos\theta + F_{1,3} = 0 \qquad\qquad F_{1,2} \cdot \sin\theta + (5/2) \cdot W = 0$$

where the $\tan\theta = \alpha$ and given $\alpha = \sqrt{3}$ so $\sin\theta = \sqrt{3}/2$ and $\cos\theta = 1/2$. These yield

$$F_{1,2} = -(5/\sqrt{3}) \cdot W \qquad\qquad F_{1,3} = (5/2\sqrt{3}) \cdot W$$

The negative sign indicates that member 1-2 is in compression.

4. *Repeat the previous step in the procedure.*

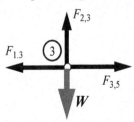

Having found the forces in members 1,2 and 1,3, node, or joint, # 3 becomes a candidate for isolation.

It shows but two unknown member forces intersecting at the node. Node #12 remains a possibility as well. I choose node #3. Force equilibrium yields

$$-F_{1,3} + F_{3,5} = 0 \qquad \text{and} \qquad +F_{2,3} - W = 0$$

Note how on the isolation I have, according to convention, assumed all member forces positive in tension. $F_{1,3}$ acts to the left, pulling on pin # 3. *This* force vector is the equal and opposite, internal reaction to the $F_{1,3}$ shown in the isolation of node # 1. With $F_{1,3} = (5/2\sqrt{3})$W   we have

$$F_{3,5} = (5/2\sqrt{3}) \cdot W \qquad\qquad \text{and} \qquad\qquad +F_{2,3} = W$$

These equations are thus, easily solved, and we go again, choosing either node # 2 or # 12 to isolate in the next step.

    5. Stopping rule: Stop when all member forces have been determined.

This piece of machinery is called *the method of joints*. Statically determinate truss member forces can be produced using other, just as sure-fire, procedures.(See problem 3.15) The main point to note is that all the member forces in a truss can be determined *from equilibrium conditions alone* using a judiciously chosen sequence of isolations of the nodes **if and only if** the truss is statically determinate. That's a circular statement if there ever was one, but you get the point[4].

3.2   Internal Forces and Moments in Beams

A *beam* is a structural element like the truss member but, unlike the latter, it is designed, fabricated, and assembled to carry a load in *bending* [5]. In this section we will go as far as we can go with our current vocabulary of force, couple, and moment and with our requirements of static equilibrium, attempting to explain what bending is, how a beam works, and even when it might fail.

The Cantilever according to Galileo

You, no doubt, know what a beam is in some sense, at least in some ordinary, everyday sense. Beams have been in use for a long time; indeed, there were beams before there were two-force members. The figure below shows a seventeenth century *cantilever beam*. It appears in a book written by Galileo, his *Dialogue Concerning Two New Sciences*.

---

4. Note how, if I were to add a redundant member connecting node #3 to node #4, I could no longer find the forces in the members joined at node #3 (nor those in the members joined at nodes #2 and #5). The problem would become equilibrium indeterminate.
5. Here is another circular statement illustrating the difficulty encountered in writing a dictionary which must necessarily turn in on itself.

Galileo wanted to know when the cantilever beam would break. He asked: What weight, hung from the end of the beam at *C*, would cause failure?

You might wonder about Galileo's state of mind when he posed the question. From the looks of the wall it is the latter whose failure he should be concerned with, not the beam. No. You are reading the figure incorrectly; you need to put on another special pair of eyeglasses that filter out the shrubbery and the decaying wall and allow you to see only a cantilever beam, rigidly attached to a rigid support at the end *AB*. These glasses will also be necessary in what follows, so keep them on.

Galileo had, earlier in his book, discussed the failure of what we would call a bar in uniaxial tension. In particular, he claimed and argued that the tensile force required for failure is proportional to the cross sectional area of the bar, just as we have done. We called the ratio of force to

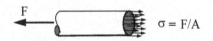

$\sigma = F/A$

area a "stress". Galileo did not use our language but he grasped, indeed, might be said to have invented the concept, at least with respect to this one very important trait – stress as a criterion for failure of a bar in tension. Galileo's achievement in analyzing the *cantilever beam under an end load* lay in relating the end load at failure to the failure load of a bar in uniaxial tension. Of course the bar had to be made of the same material. His analysis went as follows:

He imagined the beam to be an angular lever pivoted at *B*. The weight, *W*, was suspended at one end of the lever, at the end of the long arm *BC*. A horizontally directed, internal, tensile force - let us call it F_{AB} - acted along the other shorter, vertical arm of the lever *AB*. Galileo claimed this force acted at a point half way up the lever arm and provided the internal resistance to fracture.

Look back at Galileo's figure with your special glasses on. Focus on the beam. See now the internal resistance acting along a plane cut through the beam at *AB*. Forget the possibility of the wall loosening up at the root of the cantilever. Take a peek ahead at the next more modern figure if you are having trouble seeing the internal force resultant acting on the section *AB*.

For moment equilibrium about the point B one must have

$$(h/2)F_{AB} = W \cdot L$$

where I have set h equal to the height of the beam, AB, and L equal to the length of the beam, BC.

According to Galileo, the beam will fail when the ratio of F_{AB} to the cross sectional area reaches a particular, material specific value[6]. This ratio is what we have called the failure stress in tension. From the above equation we see that, for members with the same cross section area, the end load, W, to cause failure of the member acting as a cantilever is much less than the load, F_{AB}, which causes failure of the member when loaded axially, as a truss member (by the factor of $(1/2)h/L$).

A more general result, for beams of rectangular cross section but different dimensions, is obtained if we express the end load at failure in terms of the failure stress in tension, i.e., σ_{failure}:

$$W_{failure} = \frac{1}{2}(h/L) \cdot bh \cdot \sigma_{failure} \qquad \text{where} \qquad \sigma_{failure} = F_{AB}\big|_{failure} / (bh)$$

and where I have introduced b for the breadth of the beam. Observe:

- This is a quite general result. If one has determined the value of the ratio σ_{failure} for a specimen in tension, what we would call the failure stress in a tension test, then this one number provides, inserting it into the equation above, a way to compute the end load a cantilever beam, of arbitrary dimensions h, b and L, will support before failure.

- Galileo has done all of this without drawing an isolation, or free-body diagram!

- He is wrong, precisely **because he did not draw an isolation**[7].

To state he was wrong is a bit too strong. As we shall see, his achievement is real; he identified the underlying form of beam bending and its resistance to fracture. Let us see how far we can proceed by drawing an isolation and attempting to accommodate Galileo's story.

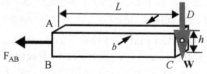

---

6. Galileo mentions wood, glass, and other materials as possibilities.
7. This claim is a bit unfair and philosophically suspect: The language of mechanics was little developed at the dawn of the 17th century. "Free body diagram" was not in the vocabulary.

I have isolated the cantilever, cutting it at *AB* away from the rest of the beam nested in the wall. Here is where Galileo claims fracture will occur. I have shown the weight *W* at the end of the beam, acting downward. I have neglected the weight of the material out of which the beam itself is fabricated. Galileo did the same and even described how you could take the weight into account if desired. I have shown a force F_{AB}, the internal resistance, acting halfway up the distance *AB*.

Is this system in equilibrium? No. Force equilibrium is not satisfied and moment equilibrium about any other point but *B* is not satisfied. This is a consequence of the failure to satisfy force equilibrium. That is why he is wrong.

On the other hand, we honor his achievement. To see why, let us do our own isolation, and see how far we can go using the static equilibrium language skills we have learned to date.

We allow that there may exist at the *root of the cantilever*, at our cut *AB*, a force, F_V and a couple M_0. We show only a vertical component of the internal

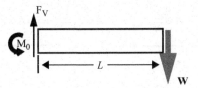

reaction force since if there were any horizontal component, force equilibrium in the horizontal direction would not be satisfied. I show the couple acting positive counter clockwise, i.e., directed out of the plane of the paper.

Force equilibrium then yields

$$F_V - W = 0 \qquad \text{or} \qquad F_V = W$$

and moment equilibrium

$$M_0 - W \cdot L = 0 \qquad \text{or} \qquad M_0 = WL$$

And this is as far as we can go; we can solve for the vertical component of the reaction force at the root, F_V, and for the couple (as we did in a prior exercise), M_0, and that's it. But notice what has happened: There is no longer any horizontal force F_{AB} to compare to the value obtained in a tension test!

It appears we (and Galileo) are in serious trouble if our intent is to estimate when the beam will fail. Indeed, we can go no further.[8] This is as far as we can go with the requirements of static equilibrium.

---

8. That is, if our criterion for failure is stated in terms of a maximum tensile (or compressive) stress, we can not say when the beam would fail. If our failure criterion was stated in terms of maximum bending moment, we *could* say when the beam would fail. But this would be a very special rule, applicable only for beams with identical cross sections and of the same material.

Before pressing further with the beam, we consider another problem, — a truss structure much like those cantilevered crane arms you see operating in cities, raising steel and concrete in the construction of many storied buildings. We pose the following problem.

Exercise 3.4

Show that truss member AC carries a tensile load of 8W, the diagonal member BC a compressive load of √2 W, and member BD a compressive load of 7W. Then show that these three forces are equivalent to a vertical force of magnitude W and a couple directed counter clockwise of magnitude WL.

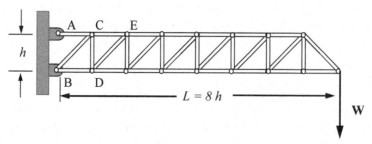

We could, at this point, embark on a *method of joints*, working our way from the right-most node, from which the weight W is suspended, to the left, node by node, until we reach the two nodes at the support pins at the wall. We will not adopt that time consuming procedure but take a short cut. We cut the structure away from the supports at the wall, just to the right of the points A and B, and construct the isolation shown below:

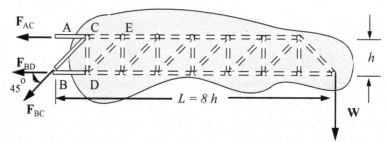

The diagram shows that I have taken the unknown, member forces to be positive in tension; F_{AC} and F_{BC} are shown pulling on node C and F_{BD} pulling on node D according to my usual convention. Force equilibrium in the horizontal and vertical directions respectively gives

$$-F_{AC} - (\sqrt{2}/2) \cdot F_{BC} - F_{BD} = 0 \qquad \text{and} \qquad -(\sqrt{2}/2)F_{BC} - W = 0$$

while moment equilibrium about point B, taking counterclockwise as positive yields

$$h \cdot F_{AC} \big| - (8h) \cdot W = 0$$

Solution produces the required result, namely

$$F_{AC} = 8W; \qquad F_{BC} = -\sqrt{2}\, W; \qquad F_{BD} = -7W$$

The negative sign in the result for F_{BC} means that the internal force is oppositely directed from what was assumed in drawing the free-body diagram; the member is in *compression* rather than tension. So too for member BD; it is also in compression. The three member forces are shown compressive or tensile according to the solution, in the isolation below, at the left. In the middle we show a statically equivalent system, having resolved the compressive force in *BC* into a vertical component, magnitude W, and a horizontal component magnitude W, then summing the latter with the horizontal force $7W$. On the right we show a statically equivalent system acting at the same section, AB – a vertical force of magnitude W and a couple of magnitude $8W\, h = W\, L$ directed counter clockwise.

Observe:

- The identity of this truss structure with the cantilever beam of Galileo is to be noted, i.e., how the moment of the weight W about the point B is balanced by the couple WL acting at the section AB. The two equal and opposite forces of magnitude 8W separated by the distance $h = L/8$ are equivalent to the couple WL.

- The most important member forces, those largest in magnitude, are the two members AC and BD. The top member AC is in tension, carrying 8W, the bottom member BD in compression, carrying 7W. The load in the diagonal member is relatively small in magnitude; it carries $1.4W$ in compression.

- Note that if I were to add more *bays* to the structure, extending the truss out to the right from *8h* to *10h*, to even *100h*, the tension and compression in the top and bottom members grow accordingly and approach the same magnitude. If $L = 100h$, then $F_{AC} = 100W$, $F_{BD} = 99W$, while the force in the diagonal member is, as before, *1.4W* in compression! Its magnitude relative to the aforementioned tension and compression becomes less and less.

We faulted Galileo for not recognizing that there must be a vertical, reaction force at the root of the cantilever. We see now that maybe he just ignored it because he knew from his (faulty)[9] analysis that it was small relative to the internal forces acting normal to the cross section at *AB*. Here is his achievement: he saw that the mechanism responsible for providing resistance to bending within a beam is the tension (and compression) of its longitudinal fibers.

Exercise 3.5

A force per unit area, a stress σ, acts over the cross section AB as shown below. It is horizontally directed and varies with vertical position on AB according to

$$\sigma(y) = c \cdot y^n \qquad\qquad -(h/2) \le y \le (h/2)$$

In this, c is a constant and n a positive integer.
If the exponent *n* is odd *show that*

(*a*) this stress distribution is equivalent to a couple alone (no resultant force), and

(*b*) the constant *c*, in terms of the couple, say M_0, may be expressed as

$$c = (n+2) \cdot M_0 / [2b \cdot (h/2)^{n+2}]$$

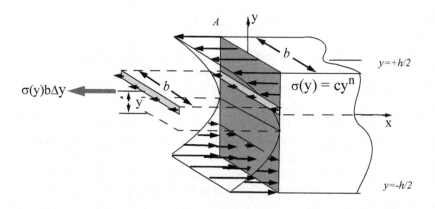

---

9.   We see how the question of evaluating Galileo's work as correct or faulty becomes complex once we move beyond the usual text-book, hagiographic citation and try to understand what he actually did using his writings as a primary source. See Kuhn, THE STRUCTURE OF SCIENTIFIC REVOLUTIONS, for more on the distortion of history at the hands of the authors of text-books in science and engineering.

First, the *resultant force*: A *differential element of force*, $\Delta F = \sigma(y)b\Delta y$ acts on each differential element of the cross section AB between the limits $y = \pm\ h/2$. Note the dimensions of the quantities on the right: σ is a force per unit area; b a length and so too Δy; their product then is a force alone. The *resultant force*, F, is the sum of all these differential elements of force, hence

$$F = \int_{-h/2}^{h/2} \sigma(y)b\,dy = c \int_{-h/2}^{h/2} y^n \cdot b\,dy$$

If the exponent n is odd, we are presented with the integral of an odd function, $-\sigma(y) = \sigma(-y)$, between symmetric limits. The sum, in this case, must be zero. Hence the resultant force is zero.

The resultant moment is obtained by summing up all the differential elements of moment due to the differential elements of force. The resultant moment will be a couple; indeed, it can be pictured as the sum of the couples due to a differential element of force acting at $+y$ and a paired differential element of force, oppositely directed, acting at $-y$. We can write, as long as n is odd

$$M_0 = 2\int_0^{h/2} y\cdot\sigma(y)b\,dy = 2c\int_0^{h/2} y^{n+1}\cdot b\,dy$$

Carrying out the integration, we obtain

$$M_0 = \frac{2cb}{(n+2)}\cdot(h/2)^{n+2}$$

So c can be expressed in terms of M_0 as

$$c = (n+2)\cdot M_0/[2b\cdot(h/2)^{n+2}]$$

as we were asked to show.

Now we imagine the section AB to be a section at the root of Galileo's cantilever. We might then, following Galileo, claim that if the maximum value of this stress, which is engendered at $y = +\ h/2$, reaches the failure stress in a tension test then the cantilever will fail. At the top of the beam the maximum stress expressed in terms of M_0 is found to be, using our result for c,

$$\sigma(y{=}h/2) = 2(n+2)\cdot M_0/(bh^2)$$

Now observe:

- The dimensions are correct: Sigma, a stress, is a force per unit area. The dimensions of the right hand side are the same - the ratio of force to length squared.

- There are many possible odd values of n each of which will give a different value for the maximum stress σ at the top of the beam. The problem, in short, is *statically indeterminate*. We cannot define a unique stress distribution satisfying moment equilibrium nor conclude when the beam will fail.

- If we **arbitrarily choose** $n = 1$, i.e., a *linear distribution of stress* across the cross section *AB*, and set $M_0 = WL$, the moment at the root of an end-loaded cantilever, we find that the maximum stress at $y = h/2$ is

$$\sigma|_{max} = 6 \cdot (L/h) \cdot (W/bh)$$

- Note the factor L/h: As we increase the ratio of length to depth while holding the cross sectional area, bh, constant — say (L/h) increases from 8 to 10 or even to 100 — the maximum stress is magnified accordingly. This "levering action" of the beam in bending holds for other values of the exponent n as well! We must credit Galileo with seeing the cantilever beam as an angular lever. Perhaps the deficiency of his analysis is rooted in his not being conversant with the concept of couple, just as students learning engineering mechanics today, four hundred years later, will err in their analyses, unable, or unwilling, to grapple with, and appropriate for their own use, the moment due to two, or many pairs of, equal and opposite forces as a thing in itself.

- If we compare this result with what Galileo obtained, identifying $\sigma_{maximum}$ above with $\sigma_{failure}$ of the member in tension, we have a factor of 6 where Galileo shows a factor of 2. That is, from the last equation, we solve for W with $\sigma_{maximum} = \sigma_{failure}$ and find

$$W_{failure} = \frac{1}{6}(h/L) \cdot bh \cdot \sigma_{failure}$$

- The beam is a redundant structure in the sense that we can take material out of the beam and still be left with a coherent and usuable structure. For example, we might *mill* away material, cutting into the sides, the whole length of the beam as shown below and still be left with a stable and possibly more *efficient* structure —A beam requiring less material, hence less cost, yet able to support the design loads.

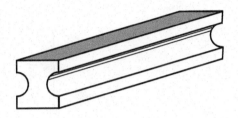

Exercise 3.6

The cross section of an I beam looks like an "I". The top and bottom parts of the "I" are called the flanges; the vertical, middle part is called the web.

If you assume that:

i) the web carries no load, no normal stress

ii) a uniformly distributed normal stress is carried by the top flange

iii) a uniformly distributed normal stress is carried by the bottom flange

iv) the top and bottom flanges have equal cross sectional areas.

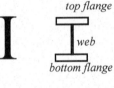

then show that

a) the resultant force, acting in the direction of the length of the beam is zero only if the stress is tensile in one of the flanges and compressive in the other and they are equal in magnitude;

b) in this case, the resultant moment, about an axis perpendicular to the web, is given by

$$M_0 = h \cdot (bt) \cdot \sigma$$

where h is the height of the cross section, b the breadth of the flanges, t their thickness.

The figure at the right shows our I beam. Actually it is an abstraction of an I beam. Our I beam, with its paper thin web, unable to carry any stress, would fail immediately.[10]

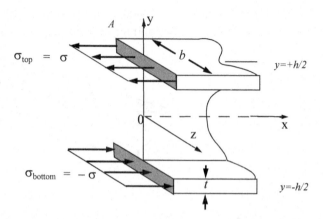

But our abstraction is not useless; it is an approximation to the way an *I* beam carries a load in bending. Furthermore, it is a *conservative* approximation in the sense that if the web *does* help carry the load (as it does), then the stress levels we obtain from our analysis, our model, should be greater than those seen by the flanges in practice.

In a sense, we are taking advantage of the indeterminacy of the problem — the problem of determining the stress distribution over the cross section of a beam in terms of the applied loading — to get some estimate of the stresses generated in an I beam. What we are asked to show in a) and b) is that the requirements of static equilibrium may be satisfied by this assumed stress distribution. (We don't worry at this point, about force equilibrium in the vertical direction).

The figure shows the top flange in tension and the bottom in compression. According to the usual convention, we take a tensile stress as positive, a compressive stress as negative. It should be clear that there is no resultant force in the horizontal direction given the conditions i) through iv). That is, force equilibrium in the (negative) x direction yields

$$\sigma_{top} \cdot (bt) + \sigma_{bottom} \cdot (bt) = 0 \qquad \text{if} \qquad \sigma_{bottom} = -\sigma_{top}$$

The resultant moment is *not* zero. The resultant moment about the 0z axis, taking them counter clockwise, is just

$$M_0 = \sigma_{top} \cdot bt \cdot (h/2) - \sigma_{bottom} \cdot bt \cdot (h/2) = 2\sigma_{top} \cdot bt \cdot (h/2) = \sigma \cdot bth$$

where I have set $\sigma_{top} = \sigma$ and $\sigma_{bottom} = -\sigma$.

With this result, we can estimate the maximum stresses in the top and bottom flanges of an I beam. We can write, if we think of M_0 as balancing the end load W of our cantilever of length L so that we can set $M_0 = WL$, and obtain:

$$\sigma_{max} = (L/h) \cdot (W/bt)$$

This should be compared with results obtained earlier for a beam with a rectangular cross section.

---

10. No I beam would be fabricated with the right-angled, sharp, interior corners shown in the figure; besides being costly, such features might, depending upon how the beam is loaded, engender *stress concentrations* — high local stress levels.

We can not resolve the indeterminacy of the problem and determine when an I beam, or any beam for that matter, will fail until we can pin down just what normal stress distribution over the cross section is produced by an internal moment. For this we must consider the deformation of the beam, how the beam deforms due to the internal forces and moments.

Before going on to that topic, we will find it useful to pursue the behavior of beams further and explore how the *shear force* and *bending moment* change with position along a span. Knowing these internal forces and moments will be prerequisite to evaluating internal stresses acting at any point within a beam.

Shear Force and Bending Moment in Beams

Indeed, we will be bold and state straight out, as conjecture informed by our study of Galileo's work, that failure of a beam in bending will be due to an excessive bending moment. Our task then, when confronted with a beam, is to determine *the bending moment distribution* that is, how it varies along the span so that we can ascertain the section where the maximum bending moment occurs.

But first, a necessary digression to discuss sign conventions as they apply to internal stresses, internal forces, and internal moments. I reconsider the case of a bar in uniaxial tension but now allow the internal stress to vary along the bar. A uniform, solid bar of rectangular cross section, suspended from above and hanging vertically, loaded by its own weight will serve as a vehicle for explanation.

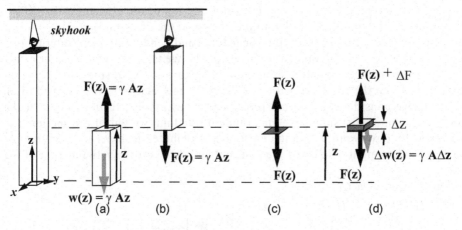

The section shown at (a) is a true free body diagram of a portion of the bar: the section has length "z", so in that sense it is of arbitrary length. The section experiences a gravitational force acting vertically downward; its magnitude is given by the product of the weight density of the section, γ, say in pounds per cubic inch, and the volume of the section which, in turn, is equal to the product of the cross sectional area, A, and the length, z. At the top of the section, where it has been "cut" away from the rest above, an internal, tensile force acts which, if force equilibrium is to be satisfied, must be equal to the weight of the section, w(z). **By convention we say that this force, a tensile force, is positive**.

The section of the bar shown at (b) is not a true free body diagram since it is not cut free of all supports (and the force due to gravity, acting on the section, is not shown). But what it *does* show is the "equal and opposite reaction" to the force acting internally at the cut section, F(z).

The section of the bar show at (c) is *infinitely thin*. It too is in tension. We speak of the tensile force at the point of the cut, at the distance z from the free end. What at first glance appear to be two forces acting at the section — one directed upward, the other downward — are, in fact, one and the same single internal force. They are both positive and have the same magnitude.

To claim that these two oppositely directed forces are the same force can create confusion in the minds of those unschooled in the business of equal and opposite reactions; but that's precisely what they are. The best way to avoid confusion is to include in the definition of the direction of a positive internal force, some specification of the surface upon which the force acts, best fixed by the direction of *the outward normal* to the surface. This we will do. In defining a positive truss member force, we say the force is positive if it acts on a surface whose outward pointing normal is in the same direction as the force acting on the surface. The force shown above is then a positive internal force — a tension.

The section shown at (d) is a *differential section (or element)*. Here the same tensile force acts at z (directed downward) but it is *not* equal in magnitude to the tensile force acting at z+Δz, acting upward at the top of the element. The difference between the two forces is due to the weight of the element, Δw(z).

To establish a convention for the shear force and bending moment internal to a beam, we take a similar approach. As an example, we take our now familiar cantilever beam an make an isolation of a section of span starting at some arbitrary distance *x* out from the root and ending at the right end, at *x* = *L*. But instead of an end load, we consider the internal forces and moments due to the weight of the beam itself. Figure (a) shows the magnitude of the total weight of the section acting vertically downward due to the uniformly distributed load per unit length, γA, where γ is the weight density of the material and *A* the cross-sectional area of the beam.

The section is a true free body diagram of a portion of the beam: the section has length L-x, so in that sense it is of arbitrary length. At the left of the section, where it has been "cut" away from the rest of the beam which is

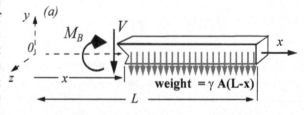

attached to the wall, we show an internal force and (bending) moment at *x*. We take it as a convention, one that we will adhere to throughout the remainder of this text, that **the shear force and the bending moment are positive as shown.** We designate the shear force by V, following tradition, and the bending moment by M_b.

Now this particular convention requires elaboration: First consider the rest of the cantilever beam that we cut away. Figure (b) shows the equal and opposite reactions to the internal force and moment shown on our free body diagram in figure (a). (b) is not a true free body diagram since it is not cut free of all supports and the force due to gravity is not shown.

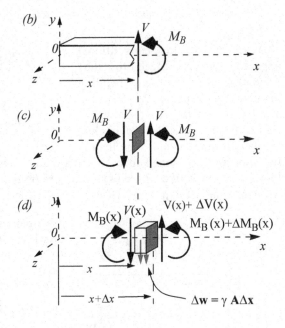

The section of the beam shown at (c) is infinitely thin. Here, what appears to be two forces is in fact one and the same internal force — the shear force, V, acting at the section x. They are both positive and have the same magnitude. Similarly what appears to be two moments is in fact one and the same internal moment — the bending moment, M_B, acting at the position x.

We show a positive shear force acting on the left face, a face with an outward normal pointing in the **negative** x direction, acting downward in the a **negative** y direction. It's equal and opposite reaction, the **same** shear force, is shown acting on the right face, a face with an outward normal pointing in the **positive** x direction, acting upward in a **positive** y direction. Our convention can then be stated as follows: **A positive shear force acts on a positive face in a positive coordinate direction or on a negative face in a negative coordinate direction.**

A **positive face** is short for **a face whose outward normal is in a positive coordinate direction**. The convention for positive bending moment is the same but now the direction of the moment is specified according to the right hand rule. We see that on the positive x face, the bending moment is positive if it is directed along the positive z axis. **A positive bending moment acts on a positive face in a positive coordinate direction or on a negative face in a negative coordinate direction.** *Warning*: Other textbooks use other conventions. It's best to indicate your convention on all exercises, including in your graphical displays the sketch to the right.

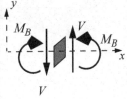

Exercise 3.7

Construct a graph that shows how the bending moment varies with distance along the end-loaded, cantilever beam. Construct another that shows how the internal, transverse shear force acting on any transverse section, varies.

With all of this conventional apparatus, we can proceed to determine the shear force and bending moment which act internally at the section x along the end-loaded cantilever beam. In this, we neglect the weight of the beam. The load at

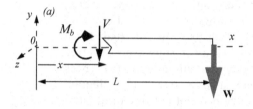

the end, W, is assumed to be much greater. Otherwise, our free body diagram looks very much like figure (a) on the previous page: Force equilibrium gives but one equation $-V - W = 0$

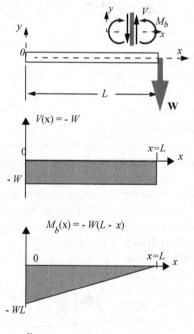

while moment equilibrium, taken about a point anywhere along the section at x gives, assuming a couple or moment is positive if it tends to rotate the isolated body counterclockwise

$$-M_b - W \cdot (L - x) = 0$$

The shear force is then a constant; it does not vary as we move along the beam, while the bending moment varies linearly with position along the beam, i.e.,

$$V = -W$$

$$\text{and}$$

$$M_b = -W \cdot (L - x)$$

. These two functions are plotted at the right, along with a sketch of the endloaded cantilever; these are the required constructions.

Some observations are in order:

• The shear force is constant and equal to the end load W but it is negative according to our convention.

• The maximum bending moment occurs at the root of the cantilever, at $x=0$; this is

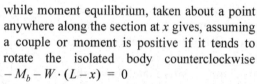

where failure is most likely to occur, as Galileo was keen to see. It too is negative according to our convention.

- The shear force is the negative of the slope of the bending moment distribution. That is

- $$V(x) = - dM_b(x)/dx$$

- If, instead of isolating a portion of the beam to the **right** of the station x,
we had isolated the portion to the **left** of the station x, we could have solved the
problem but we would have had to have first evaluated the reactions at the wall.

- The isolation shown at the right and the application of force and moment
equilibrium produce the same shear force and bending moment distribution as
above. Note that the reactions shown at the wall, at $x=0$, are displayed according to their true directions; they can be considered the applied forces for this
alternate, free body diagram.

Exercise 3.8

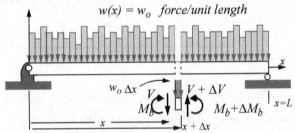

Show that for the uniformly loaded beam simply supported at its ends, the
following differential relationships among the distributed load w_0, the
shear force $V(x)$, and the bending moment $M_b(x)$, hold true, namely

$$\frac{dV}{dx} = w_0 \qquad \text{and} \qquad \frac{d}{dx}M_b = -V$$

The differential relations among the shear force,
$V(x)$, the bending moment, $M_b(x)$ and the distributed
load w_0 are obtained from imagining a short, differential element of the beam of length Δx, cut out from the
beam at some distance x. In this particular problem we
are given a uniformly distributed load. Our derivation,

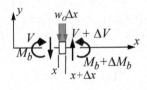

however, goes through in the same way if w_0 is not constant but varies with x, the
distance along the span. The relationship between the shear force and $w(x)$ would
be the same.

Such an element is shown above. **Note the difference between this differen-**
tial element sketched here and the pictures drawn in defining a convention for

positive shear force and bending moment: They are alike but they are to be read differently. The sketch used in defining our convention shows the internal force and moment at **a point** along the span of the beam; the sketch above and in (d) shows how the internal force and moment change over a small, but finite, length of span – over **a differential element**.

Focusing on the isolation of this differential element of the beam, force equilibrium requires

$$-V - w_o \cdot \Delta x + (V + \Delta V) = 0$$

and moment equilibrium, about the point x, counter clockwise positive, yields

$$- M_b(x) - w_0 \cdot \Delta x \cdot (\Delta x /2) + (V + \Delta V) \cdot \Delta x + M_b + \Delta M_b = 0$$

We simplify, divide by Δx, let Δx approach zero and obtain for the ratios $\Delta V/ \Delta x$ *and* $\Delta M_b / \Delta x$ in the limit

$$\frac{dV}{dx} = w_0 \qquad \text{and} \qquad \frac{d}{dx} M_b = -V$$

as was desired.

Note how, because the factor Δx appears twice in the w_0 term in the equation of moment equilibrium, it *drops out* upon going to the limit. We say it is *second order* relative to the other *leading order* terms which contain but a single factor Δx. The latter are *leading order* after we have canceled out the M_b, - M_b terms. Knowing well the sign convention for positive shear force and bending moment is critical to making a correct reading of these differential equations. These general equations themselves — again, w_0 could be a function of x, *w(x)*, and our derivation would remain the same —s can be extremely useful in constucting shear force and bending moment distributions. That's why I've placed a box around them.

For example we might attempt to construct the shear force and bending moment distributions by seeking integrals for these two, first order, differential equations. We would obtain, since w_0 is a constant

$$V(x) = w_0 \cdot x + C_1 \qquad \text{and} \qquad M_b(x) = w_0 \cdot (x^2 /2) + C_1 \cdot x + C_2$$

But how to evaluate the two constants of integration? To do so we must know values for the shear force and bending moment at some x position, or positions, along the span.

Now, for our particular situation, we must have the bending moment vanish at the ends of the beam since there they are *simply supported* — that is, the supports offer no resistance to rotation hence the internal moments at the ends must be zero. This is best shown by an isolation in the vicinity of one of the two ends.

V

$M_b = 0$

$Reaction = \quad w_0\,L/2$

We require, then, that the following two *boundary conditions* be satisfied, namely

$$\text{at } x=0, \quad M_b = 0 \quad \text{and at } x=L, \quad M_b = 0$$

These two yield the following expressions for the two constants of integration, C_1 and C_2.

$$C_1 = -w_0 \cdot (L/2) \qquad \text{and} \qquad C_2 = 0$$

and our results for the shear force and bending moment distributions become:

$$V(x) = w_0 \cdot (x - L/2)$$

$$M_b(x) = \frac{w_0 L^2}{2} \cdot [(x/L)^2 - (x/L)]$$

Unfortunately, this way of determining the shear force and bending moment distributions within a beam does not work so well when one is confronted with concentrated, point loads or segments of distributed loads. In fact, while it works fine for a continuous, distributed load over the full span of a beam, as is the case here, evaluating the constants of integration becomes cumbersome in most other cases. Why this is so will be explored a bit further on.

Given this, best practice is to determine the shear force and bending moment distributions from an isolation, or sequence of isolations, of portion of the beam. The differential relationships then provide a useful check on our work. Here is how to proceed:

We first determine the reactions at the supports at the left and right ends of the span.

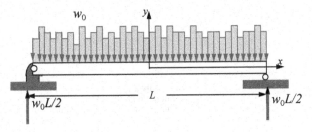

w_0

y

x

$w_0L/2$

L

$w_0L/2$

Note how I have re-positioned the axis system to take advantage of symmetry.[11]

Symmetry suggests, and a free body diagram of the entire beam together with application of force and moment equilibrium would show, that the horizontal reactions at the ends are zero and the vertical reactions are the same, namely $w_0 L/2$.

We isolate a portion of the beam to the right of some arbitrarily chosen station x. The choice of this section is not quite arbitrary: We made a cut at a positive x, a practice highly recommended to avoid sign confusions when writing out expressions for distances along the span in applying moment equilibrium.

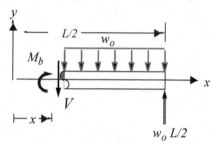

Below right, we show the same isolation but have replaced the load w_0 distributed over the portion of the span x to $L/2$, by an equivalent system, namely a force of magnitude $w_0[(L/2)-x]$ acting downward through a point located midway x to $L/2$. Applying force equilibrium to the isolation at the right yields:

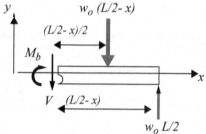

$$-V(x) - w_0 \cdot [(L/2) - x] + w_0 \cdot (L/2) = 0$$

while taking moments about the point x, counter clockwise positive, yields

$$-M_b(x) - w_0 \cdot [(L/2) - x][(L/2) - x]/2 + (w_0 L/2)[(L/2) - x] = 0$$

---

11. Note how the loading looks a bit jagged; it is not really a constant, as we move along the beam. While the effects of this "smoothing" of the applied load can not really be determined without some analysis which allows for the varying load, we note that the bending moment is obtained from an integration, twice over, of the distributed load. Integration is a smoothing operation. We explore this situation further on.

Solution of these yields the shear force and bending moment distributions shown below. We show the uniform load distribution as well.

$w(x) = w_o$

$V(x) = w_o x$

$M_b(x) = (w_0/2)[(L/2)^2 - x^2]$

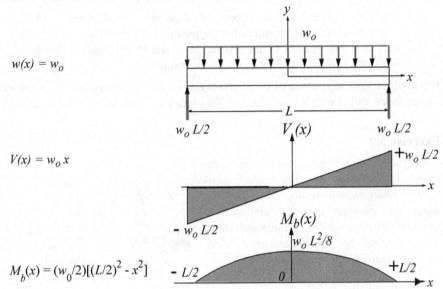

Observe:

- The bending moment is a maximum at mid-span. $M_b = w_0 L^2/8$. Note that the shear force is zero at mid-span, again in accord with our differential relationship[12].

- How by taking moments about the point x, the shear force does not appear in the moment equilibrium equation. The two equations are *uncoupled*, we can solve for $M_b(x)$ without knowing V.

- These results are the same as obtained from our solution of the differential equations. They do not immediately appear to be identical because the "x" is measured from a different position. If you make an appropriate change of coordinate, the identity will be confirmed.

- Another way to verify their consistency is to see if the differential relationships, which apply locally at any position x, are satisfied by our more recent results. Indeed they are: The **slope** of the shear force distribution is equal to the distributed load w_0 at any point x. The **slope** of the bending moment distribution is equal to the negative of the shear force $V(x)$.

12. One must be very careful in seeking maximum bending moments by setting the shear to zero. One of the disastrous consequences of studying the differential calculus is that one might think the locus of a maximum value of a function is always found by equating the slope of the function to zero. Although true in this problem, this is not always the case. If the function is discontinuous or if the maximum occurs at a boundary then the slope need not vanish yet the function may have its maximum value there. Both of these conditions are often encountered in the study of shear force and bending moment distributions within beams.

- The bending moment is zero at both ends of the span. This confirms our reading of circles as frictionless pins, unable to transmit a couple.

- Last, but not least, the units check. For example, a bending moment has the dimensions FL, force times length; the distributed load has dimensions F/L, force per unit length; the product of w_0 and L^2 then has the dimensions of a bending moment as we have obtained.

For another look at the use of the differential relationships as aids to constructing shear force and bending moment distributions we consider a second exercise:

Exercise 3.9

Construct shear force and bending moment diagrams for the simply-supported beam shown below. How do your diagrams change as the distance a approaches zero while, at the same time, the resultant of the distributed load, $w_0(x)$ remains finite and equal to P?

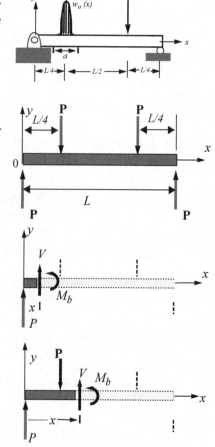

We start with the limiting case of a *concentrated load* acting at the point to the left of center span. Two isolations of portions of the beam to the left are made at some arbitrary x – first with x less than $L/4$, (middle figure), then in the region $L/4<x<3L/4$, (bottom figure)– are shown.

Symmetry again requires that the vertical reactions are equal and of magnitude P. Note this remains true when we consider the distributed load $w_0(x)$ centered at $x= L/4$ as long as its resultant is equivalent to the concentrated load P.

Force and moment equilibrium for $0< x < L/4$ yields

$$V(x) = -P$$

and

$$M_b(x) = P \cdot x`$$

while for $L/4 < x < 3L/4$ we have

$$V(x) = P - P = 0$$

$$and$$

$$M_b(x) = P \cdot x - P \cdot (x - L/4) = P \cdot L/4$$

Now for the $x > 3L/4$ we could proceed by making a third isolation, setting $x > 3L/4$ but rather than pursue that tack, we step back and construct the behavior of the shear force and bending moment in this region using less machine-like, but just as rigorous language, knowing the behavior at the end points and the differential relations among shear force, bending moment, and distributed load.

The distributed load is zero for $x > 3L/4$. Hence the shear force must be a constant. But what constant value? We know that the reaction at the right end of the beam is P acting upward. Imagining an isolation of a small segment of the beam at $x \approx L$, you see that the shear force must equal a positive P. I show the convention icon at the right to help you imagine the a true isolation at x=L.

In the region, $3L/4 < x < L$ we have,

then $\qquad V(x) = P$

For the bending moment in this region we can claim that if the shear force is constant, then the bending moment must be a linear function of x with a slope equal to $-V$, i.e., $= -P$. The bending moment must then have the form

$$M(x) = -P \cdot x + C$$

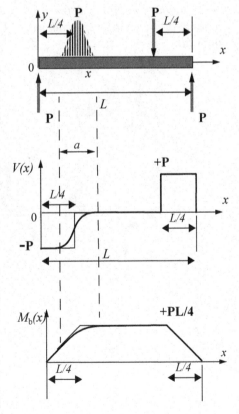

where C is a constant. But the bending moment at the right end is zero. From this we can evaluate C, conclude that the bending moment is a straight line, zero at x=L and with slope equal to -P, i.e., it has the form: $\qquad M_b(x) = P \cdot (L - x)$

I have also indicated the effect of distributing the load P out over a finite segment, a of the span, centered at x=L/4. Since the distributed P is equivalent to a $w(x)$, acting downward as positive, then the slope of the shear V must be positive according to our differential relationship relating the two. The bending moment too changes, is *smoothed* as a result, its slope, which is equal to -V, is less for x<L/4 and greater than it was for x>L/4.

We see that the effect of distributing a concentrated load is to eliminate the discontinuity, the jump, in the shear force at the point where the concentrated load is applied. We also see that the discontinuity in the **slope** of the bending moment distribution at that point dissolves.

Now while at first encounter, dealing with functions that jump around can be disconcerting, reminiscent of all of that talk in a mathematics class about limits and their existence, we will welcome them into our vocabulary. For although we know that concentrated loads are as rare as frictionless pins, like frictionless pins, they are extremely useful abstractions in engineering practice. You will learn to appreciate these rare birds; imagine what your life would be like if you had to check out the effect of friction at every joint in a truss or the effect of deviation from concentration of every concentrated load P?

One final exercise on shear force and bending moment in a beam:

Exercise 3.10

Estimate the magnitude of the maximum bending moment due to the uniform loading of the cantilever beam which is also supported at its end away from the wall.

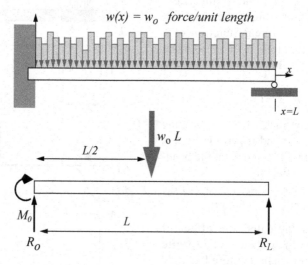

We first determine, or try to determine, the *reactions at the wall* and at the *roller support* at the right end.

Force and moment equilibrium yield,

$$R_0 - w_0 L + R_L = 0$$

$$and$$

$$-M_0 - w_0(L^2/2) + R_L \cdot L = 0$$

Here moments have been taken about left end, positive counterclockwise. Also, I have replaced the uniformly distributed load, w_0 with a statically equivalent load equal to its resultant and acting at midspan.

Now these are two equations but there are three unknown reactions, R_O, R_L, M_O. The problem is *indeterminate*, the structure is *redundant*; we could remove the support at the right end and the shelf would still work to hold up the books, assuming we do not overload the now cantilevered, structure. But with the support at the right in place, life is hard, or at least more complex.

But wait; all that was asked was an *estimate* of the maximum bending moment. Let us press on; we are not without resources. In fact, our redundant structure looks something like the previous exercise involving a uniformly loaded beam which was simply supported at **both** ends. There we found a maximum bending moment of $w_o L^2 / 8$ which acted at mid span. There! There is an estimate![13] Can we do better? Possibly. (See Problem 3.1)

We leave beam bending for now. We have made considerable progress although we have many loose ends scattered about.

- What is the nature of the stress distribution engendered by a bending moment?

- How can we do better analyzing indeterminate structures like the one above?

We will return to answer these questions and pick up the loose ends, in Chapter 8. For now we turn to two quite different structural elements – circular shafts in torsion, and thin cylinders under internal or external pressure – to see how far we can go with equilibrium alone in our search for criteria to judge, diagnose and design structures with integrity.

---

13. This is equivalent to setting the resistance to rotation at the wall, on the left, to zero.

3.3   Internal Moments in Shafts in Torsion

By now you get the picture: Structures come in different types, made of different elements, each of which must support internal forces and moments. The pin-ended elements of a truss structure can carry "uni-axial" forces of tension or compression. A beam element supports internal forces and moments - "transverse" shear forces and bending moments. (A beam can also support an axial force of tension or compression but this kind of action does not interact with the shear force and bending moment - unless we allow for relatively large displacements of the beam, which we shall do in the last chapter). We call a structure made up of beam elements a "frame".

Structural elements can also twist about their axis. Think of the drive shaft in an automobile transmission. The beam elements of a frame may also experience torsion. A shaft in torsion supports an internal moment, a torque, about its "long" axis of rotation.

Exercise 3.11

Estimate the torque in the shaft RH appearing in the figure below

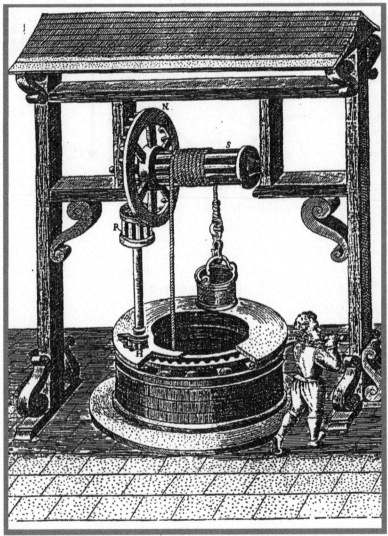

This figure, of a human-powered pump, is taken from THE VARIOUS AND INGENIOUS MACHINES OF AGOSTINO RAMELLI, a sixteenth century, late Renaissance work originally published in Italian and French.

We isolate pieces of the structure in turn, starting with the drum S upon its shaft at the top of the machine, then proceed to the vertical shaft RH to estimate the torque it bears. We assume in all of our fabrications that the bearings are frictionless, they can support no torque, they provide little resistance to rotation.[14]

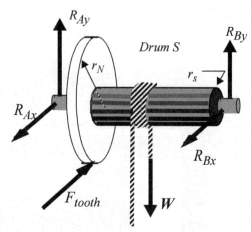

We show the reactions at the two bearings as R_A and R_B. Their values are not of interest; we need only determine the force acting on the teeth of the wheel N, labeled F_{tooth}, in order to reach our goal.

Moment equilibrium about the axis of the shaft yields

$$F_{tooth} = W \cdot (r_x/r_n)$$

where W is the weight of the water bucket, assumed full of water, r_s is the radius of the drum S, and r_N the radius of the wheel N out to where the internal force acting between the teeth of wheel N and the "rundles" of the "lantern gear" R.

We now isolate the vertical shaft, rather a top section of the vertical shaft, to expose the internal torque, which we shall label M_T. On this we show the equal and opposite reaction to the tooth force acting on the wheel N, using the same symbol F_{tooth}. We let r_R be the radius of the lantern gear. We leave for an end-of-chapter exercise the problem of determining the reaction force at the bearing (not labeled) and another at the bottom of the shaft.

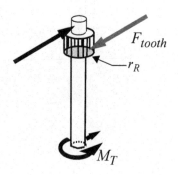

Moment equilibrium about the axis of the shaft yields

---

14. This is an adventurous assumption to make for the sixteenth century but, in the spirit of the Renaissance and Neo-platonic times, we will go ahead in this fashion. The drawings that are found in Ramelli's book are an adventure in themselves. Page after page of machinery - for milling grain, cranes for lifting, machines for dragging heavy objects without ruining your back, cofferdams, military screwjacks and hurling engines, as well as one hundred and ten plates of water-raising devices like the one shown here - can be read as a celebration of the rebirth of Western thought, and that rebirth extended to encompass technology. This, in some ways excessive display of technique – many of the machines are impractical, drawn only to show off – has its parallel in contemporary, professional engineering activity within the academies and universities. Witness the excessive production of scholarly articles in the engineering sciences whose titles read like one hundred and ten permutations on a single fundamental problem.

$$M_T = F_{tooth} \cdot r_R$$

or, with our expression for F_{tooth}

$$M_T = W(r_R \cdot r_s / r_N)$$

Now for some numbers. I take 60 *pounds* as an estimate of the weight W. I take sixty pounds because I know that a cubic foot of water weighs 62.4 *pounds* and the volume of the bucket looks to be about a cubic foot. I estimate the radius of the drum to be $r_S = 1$ *ft*, that of the wheel to be three times bigger, $r_N = 3$ *ft*, and finally the radius of the lantern gear to be $r_R = 1$ *ft*. Putting this all together produces an estimate of the torque in the shaft of

$$M_T \approx 20 \ ft.lb$$

If Ramelli were to ask, like Galileo, when the shaft *HR* might fail, he would be hard pressed to respond. The reason? Assuming that failure of the shaft is a local, or microscopic, phenomenon, he would need to know how the torque M_T estimated above is distributed over a cross section of the shaft. The alternative would be to test every shaft of a different diameter to determine the torque at which it would fail.[15]

We too, will not be able to respond at this point. Again we see that the problem of determining the stresses engendered by the torque, more specifically, the shear stress distribution over a cross section of the shaft, is *indeterminate*. Still, as we did with the beam subject to bending, let us see how far we can go.

We need, first, to introduce the notion of *shear stress*. Up to this point we have toyed with what is called a *normal stress, normal* in the sense that it acts *perpendicular* to a surface, e.g., the tensile or compressive stress in a truss member. A *shear stress* acts *parallel* to a surface.

The figure at the right shows a thin-walled tube loaded in torssion by a torque (or moment) M_T.   The bit cut out of the top of the tube is meant to show a *shear stress* τ, distributed over the thickness and acting perpendicular to the radius of the tube. It acts parallel to the surface; we say it tends to shear that surface over the one below it; the cross section rotates a bit about the axis relative to the cross sections below.

I claim that if the tube is *rotationally symmetric*, that is, its geometry and properties do not

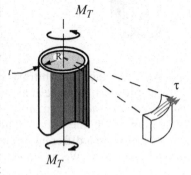

15. A torque of 20 ft-lb. is not a very big torque. The wooden shaft RH would have to be extremely defective or very slender to have a torque of this magnitude cause any problems. Failure of the shaft is unlikely. On the other hand, we might ask another sort of question at this point: What force must the worker exert to raise the bucket of water? Or, how fast must he walk round and round to deliver water at the rate of 200 gallons per hour? At this rate, how much horse power must he supply? Failure in this mode is more likely.

change as you move around the axis of the tube, then each bit of surface will look the same as that shown in the figure. Furthermore, if we assume that the shear is uniformly distributed over the thickness of the tube we can figure out how big the shear stress is in terms of the applied torque and the geometry of the tube.[16]

The contribution to the torque of an angular segment of arc length $R\Delta\theta$ will be

$\tau \cdot R\Delta\theta \cdot t$: the element of force

             times the radius

$\tau \cdot R^2\Delta\theta \cdot t$: the element of torque

so integrating around the surface of the tube gives a resultant

$2\pi R^2 t \cdot \tau$ which must equal the applied torque, hence $\tau = \dfrac{M_T}{2\pi R^2 t}$

Note that the dimensions of shear stress are force per unit area as they should be.

Exercise 3.12

Show that an equivalent system to the torque M_T acting about an axis of a solid circular shaft is a shear stress distribution $\tau(r,\theta)$ which is independent of θ but otherwise an arbitrary function of r.

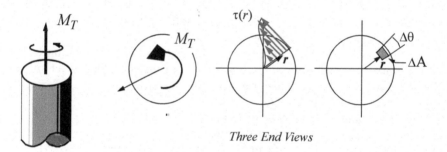

Three End Views

We show such an arbitrary shear stress τ, a force per unit area, varying from zero at the axis to some maximum value at the outer radius R. We call this a *monotonically increasing function* of r. It need not be so specialized a function but we will evaluate one of this kind in what follows.

We show too a differential element of area $\Delta A = (r\Delta\theta)(\Delta r)$, where polar coordinates are used. We assume rotational symmetry so the shear stress does not change as we move around the shaft *at the same radius*. Again, the stress distribution is *rotationally symmetric*, not a function of the polar coordinate θ. With this,

---

16. This is reminiscent of our analysis of an I beam.

for this distribution to be equivalent to the torque M_T, we must have, equating moments about the axis of the shaft:

$$M_T = \int_{Area} r \cdot [\tau \cdot \Delta A]$$

where the bracketed term is the differential element of force and r is the moment arm of each force element about the axis of the shaft.

Taking account of the rotational symmetry, summing with respect to θ introduces a factor of 2π and we are left with

$$M_T = 2\pi \int_{r=0}^{r=R} \tau(r) \cdot r^2 dr$$

where R is the radius of the shaft.

This then shows that we can construct one, or many, shear stress distribution(s) whose resultant moment about the axis of the cylinder will be equivalent to the torque, M_T. For example, we might take

$$\tau(r) = c \cdot r^n$$

where n is any integer, carry out the integration to obtain an expression for the constant c in terms of the applied torque, M_T. This is similar to the way we proceeded with the beam.

3.4   Thin Cylinder under Pressure

The members of truss structures carry the load in tension or compression. A cylinder under pressure behaves similarly in that the most significant internal force is a tension or compression. And like the truss, if the cylinder is thin, the problem of determining these internal forces is statically determinate, or at least approximately so. A few judiciously chosen isolations will enable us to estimate the tensile and compressive forces within making use, as always, of the requirements for static equilibrium. If, in addition, we assume that these internal forces are uniformly distributed over an internal area, we can estimate when the thin cylindrical shell might yield or fracture, i.e., we can calculate an internal normal stress. We put off an exploration of failure until later. We restrict our attention here to constructing estimates of the internal stresses.

Consider first an isolation that cuts the thin shell with a plane perpendicular to the cylinder's axis.

We assume that the cylinder is internally pressurized. In writing equilibrium, we take the axial force distributed around the circumference, f_a, to be uniformly distributed as it must since the problem is rotationally symmetric. Note that f_a, has dimensions *force per unit length*. For equilibrium in the vertical direction:

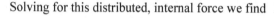

$$2\pi R \cdot f_a = p_i \cdot \pi R^2$$

Solving for this distributed, internal force we find

$$f_a = p_i \cdot (R/2)$$

If we now assume further that this force per unit length of circumference is uniformly distributed over the thickness, t, of the cylinder, akin to the way we proceeded on the thin hollow shaft in torsion, we obtain an estimate of the tensile stress, a force per unit area of the thin cross section, namely

Observe that the stress σ_a can be very much larger than the internal pressure if the ratio of thickness to radius is small. For a thin shell of the sort used in aerospace vehicles, tank trailers, or a can of coke, this ratio may be on the order of

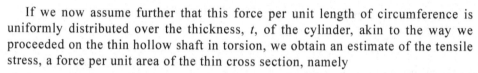

$$\sigma_a = p_i \cdot (R/2t)$$

0.01. The stress then is on the order of 50 times the internal pressure. But this is not the maximum internal normal stress! Below is a second isolation, this time of a circumferential section.

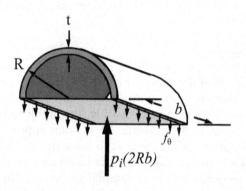

Equilibrium of this isolated body requires that

$$p_i \cdot (2Rb) = 2bf_\theta$$

where f_θ is an internal, again tensile, uniformly distributed force per unit length acting in the "theta" or *hoop* direction. Note: We do not show the pressure and the internal forces acting in the axial direction. These are self equilibrating in the sense that the tensile forces on one side balance those on the other side of the cut a distance b along the cylinder. Note also how, in writing the resultant of the internal pressure as a vertical force alone, we have put to use the results of section 2.2.

Solving, we find

$$f_\theta = p_i \cdot R$$

If we again assume that the force per unit length in the axial direction is also uniformly distributed over the thickness, we find for the *hoop stress*

$$\sigma_\theta = p_i \cdot (R/t)$$

which is twice as big as what we found for the internal stress acting internally, parallel to the shell's axis. For really thin shells, the *hoop stress* is critical.

Design Exercise 3.1

Low-end Diving Board

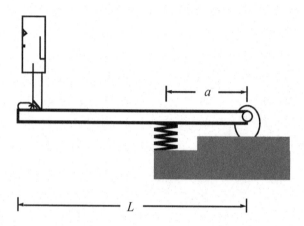

You are responsible for the design of a complete line of diving boards within a firm that markets and sells worldwide. Sketch a rudimentary design of a *generic* board. Before you start, list some performance criteria your product must satisfy. Make a list also of those elements of the diving board, taken as a whole system, that determine its performance.

Focusing on the dynamic response of the system, explore how those elements might be *sized* to give your proposed design the right *feel*. Take into account the range of sizes and masses of people that might want to make use of the board. Can you set out some criteria that must be met if the performance is to be judged good? Construct more alternative designs that would meet your main performance criteria but would do so in different ways.

Design Exercise 3.2

Low-end Shelf Bracket

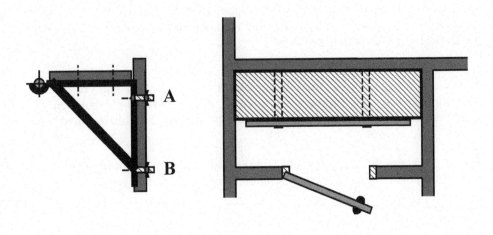

Many closets are equipped with a clothes hanger bar that is supported by two sheetmetal brackets. The brackets are supported by two fasteners *A* and *B* as shown that are somehow anchored to the wall material (1/2 inch sheetrock, for example. A shelf is then usually place on top of the brackets. There is provision to fasten the shelf to the brackets, but this is often not done. When overloaded with clothes, long-playing records, stacks of back issues of National Geographic Magazine, or last year's laundry piled high on the shelf, the system often fails by pull-out of the upper fastener at *A*.

- Estimate the pullout force acting at *A* as a function of the load on the clothes bar and shelf load.

- Given that the wall material is weak and the pullout strength at *A* cannot be increased, devise a design change that will avoid this kind of failure in this, a typical closet arrangement.

3.5   Problems - Internal Forces and Moments

3.1   *Estimate* the maximum bending moment within the tip supported, uniformly loaded cantilever of chapter exercise 3.10 using the result for a uniformly loaded cantilever which is unsupported at the right. Would you expect this to be an *upper or lower bound* on the value obtained from a full analysis of the statically indeterminate problem?

3.2   Consider the truss structure of Exercise 3.3: *What if* you are interested only in the forces acting within the members at midspan. *Show that* you can determine the forces in members 6-8, 6-9 and 7-9 with but a **single** isolation, after you have determined the reactions at the left and right ends. This is called *the method of sections.*

3.3   *Show that* for **any** exponent n in the expression for the normal stress distribution of Exercise 3.5, the maximum bending stress is given by

$$|\sigma_{max}| = 2(n+2) \cdot M_0/(bh^2)$$

If M_0 is the moment at the root of an end-loaded cantilever (end-load = W) of lenght L, then this may be written

$$|\sigma_{max}| = 2(n+2) \cdot (L/h)(W/bh)$$

hence the normal stress due to bending, for a beam with a rectangular cross section will be significantly greater than the average shear stress over the section.

3.4   *Estimate* the maximum bending moment in the wood of the clothespin shown full size. Where do you think this structure would fail?

3.5   *Construct* the shear force and bending moment diagram for Galileo's lever.

3.6   *Construct* a shear force and bending moment diagram for the truss of Exercise 3.4. Using this, estimate the forces carried by the members of the third bay out from the wall, i.e., the bay starting at node *E*.

3.7   *Construct* an expression for the bending moment at the root of the lower limbs of a mature maple tree in terms of the girth, length, number of offshoots, etc... whatever you judge important. How does the bending moment vary as you go up the tree and the limbs and shoots decrease in size and number (?).

3.8 A hand-held power drill of 1/4 horsepower begins to grab when its rotational speed slows to 120 *rpm*, that's revolutions per minute. *Estimate* the force and couple I must exert on the handle to keep a 1/4 inch drill aligned.

3.9 *Estimate* the force Ramelli's laborer (or is it Ramelli himself?) must push with in order to just lift a full bucket of water from the well shown in the figure.

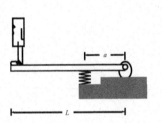

3.10 *Construct* the shear force and bending moment distribution for the diving board shown below. Assuming the board is rigid relative to the linear spring at *a*, *show that* the *equivalent stiffness* of the system at *L*, *K* in the expression $P = K\Delta$ where Δ is the deflection under the load, is

$$K = k \cdot (a/L)^2$$ where *k* is the stiffness of the linear spring at *a*.

3.11 Find the force in the member CD of the structure shown in terms of P. All members, save CF are of equal length. In this, use method of joints starting from either node B or node G, according to your teacher's instructions.

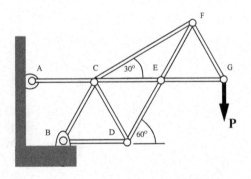

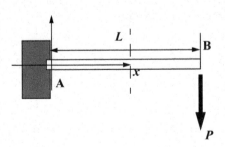

3.12 Find an expression for the internal moment and force acting at *x*, some arbitrary distance from the root of the cantilever beam. Neglect the weight of the beam.

What if you now include the weight of the beam, say w_0 per unit length; how do these expressions change?

What criteria would you use in order to safely neglect the weight of the beam?

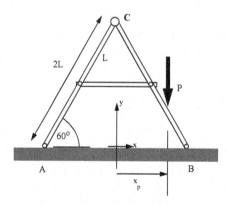

3.13 Find the reactions acting at A and B in terms of P and the dimensions shown (x_p/L).

Isolate member BC and draw a free body diagram which will enable you to determine the forces acting on this member.

Find those forces, again in terms of P and the dimensions shown.

Find the force in the horizontal member of the structure.

3.14 Determine the forces acting on member DE. How does this system differ from that of the previous problem? How is it the same?

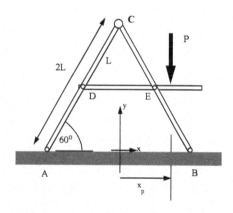

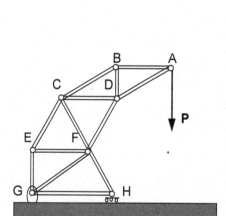

3.15 Estimate the forces acting in members EG, GF, FH in terms of P. In this, use but one free body diagram.

Note: Assume the drawing is to scale and, using a scale, introduce the relative distances you will need, in writing out the requirement of moment equilibrium, onto your free body diagram.

3.16 A simply supported beam of length L carries a concentrated load, P, at the point shown.

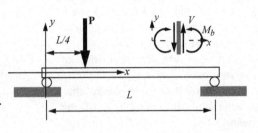

i) Determine the reactions at the supports.

ii) Draw two free body diagrams, isolating a portion of the beam to the right of the load, another to the left of the load.

iii) Apply force equilibrium and find the shear force V as a function of x over both domains Plot $V(x)$

iv) Apply moment equilibrium and find how the bending moment M_b varies with x. Plot $M_b(x)$.

v) Verify that $dM_b/dx = -V$.

3.17 A simply supported beam (indicated by the rollers at the ends) carries a trolley used to lift and transport heavy weights around within the shop. The trolley is motor powered and can move between the ends of the beam. For some arbitrary location of the trolley along the beam, a,

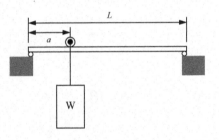

i) What are the reactions at the ends of the beam?

ii) Sketch the shear force and bending moment distributions.

iii) How does the *maximum* bending moment vary with a; i.e., change as the trolley moves from one end to the other?

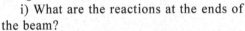
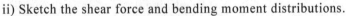
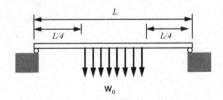

3.18 Sketch the shear force and bending moment distribution for the beam shown at the left. Where does the maximum bending moment occur and what is its magnitude.

3.19   A simply supported beam
of length L carries a uniform load
per unit length, w_0. over a portion
of the length, $\beta L < x < L$

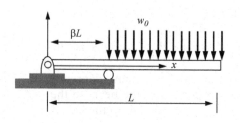

i) Determine the reactions at
the supports.

ii) Draw two free body diagrams, isolating   portions of the
beam to the right of the origin.
Note: include all relevant dimen-
sions as well as known and unknown force and moment components.

iii) Apply force equilibrium and find the shear force V as a function of x. Plot.

iv) Apply moment equilibrium and find how the bending moment M_b varies
with x. Plot.

v) Verify that $dM_b/dx \; = -V$ within each region.

3.20   Estimate the maximum bending moment within an olympic sized diving
board with a person standing at the free end, contemplating her next step.

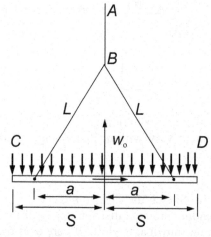

3.21   A beam, carrying a uniformly
distributed load, is suspended by cables
from the end of a crane (crane not
shown). The cables are attached to the
beam at a distance a from the center line
as shown. Given that
$$a = (3/4)S \quad and \quad L = (3/2)S$$
i) Determine the tension in the cable AB.
Express in non-dimensional form, i.e.,
with respect to $w_0 S$.

ii) Determine the tension in the cables of
length L.

iii) Sketch the beam's shear force and
bending moment diagram. Again, non-
dimensionalize. What is the magnitude
of the maximum bending moment and where does it occur?

iv) Where should the cables be attached - (a/S = ?) -to minimize the magnitude
of the maximum bending moment? What is this minimum value?

v) If a/S is chosen to minimize the magnitude of the maximum bending
moment, what then is the tension in the cables of length L? Compare with your
answer to (ii).

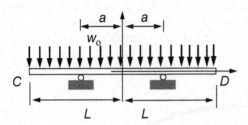

3.22 Where should the supports of the uniformly loaded beam shown at the left be placed in order to minimize the magnitude of the maximum bending moment within the beam? i.e, a/L =?

3.23 A cantilever beam with a hook at the end supports a load P as shown. The bending moment at x= 3/4 L is:

a) positive and equal to P*(L/4)

b) negative and equal to P*(3L/4)

c) zero.

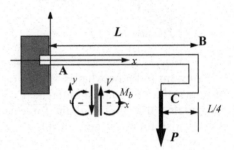

3.24 Sketch the shear force and bending moment distribution for the beam shown at the left. Where does the maximum bending moment occur and what is its magnitude.

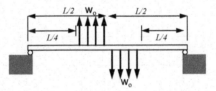

3.25 The rigid, weight-less, beam carries a load P at its right end and is supported at the left end by two (frictionless pins). What can you say about the reactions acting at A and B? e.g., "they are equivalent to..."

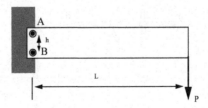

3.26   In a lab experiment, we subject a strand of pasta to an endload as show in the first figure. The strand undegoes relatively large, transverse displacement. The (uncooked) noodle bends more and more until it eventually breaks - usually at midspan - into two pieces.

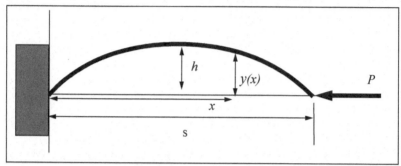

We want to know what is "going on", internally, near mid-span, before failure in terms of a force and a moment. Complete the free-body diagram begun below, recognizing that the resultant force and resultant moment on the isolated body must vanish for static equilibrium.

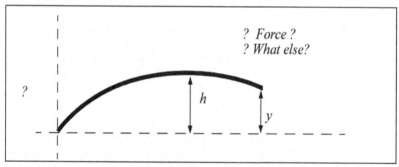

3.27   For the truss shown below,
    i) Isolate the full truss structure and replace the applied loads with an equivalent load (no moment) acting at some distance, b, from the left end. What is b?
    ii) Determine the reactions at **f** and **l**.
    iii)   Find the force in member ch with but a single additional free body diagram. (In this part, make sure you work with the external forces as originally given).

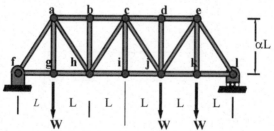

3.28   For the truss shown below,

i) Isolate the full truss structure and replace the applied loads with an equivalent load (no moment) acting at some distance, b, from the left end. What is b?

ii) Determine the reactions at **f** and **l**.

iii) Find the force in member **ch** with but a single additional free body diagram. Compare your result with that obtained in the previous problem.

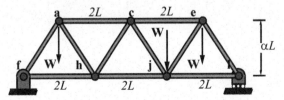

3.29   A crane, like those you encounter around MIT these days, shows a
variable geometry; the angle θ can vary from zero to almost 90 degrees and, of
course, the structure can rotate 360 degrees around the vertical, central axis of the
tower. As θ varies, the angle the heavy duty cable BC makes with the horizontal
changes and the system of pulley cables connecting C and E change in length.

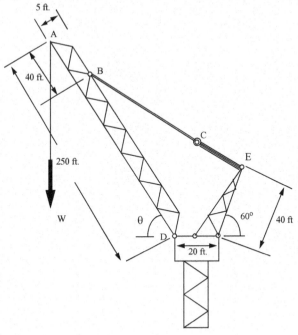

Drawing an appropriate isolation, determine both the reaction force at D, where a
(frictionless) pin connects the truss-beam to the tower, and the force, F_{BC}, in the cable BC
as functions of W and θ.

Plot F_{BC}/W as a function of θ. Note: $\theta = 60$ degrees in the configuration shown.

4

Stress

We have talked about internal forces, distributed them uniformly over an area and they became a *normal stress* acting *perpendicular* to some internal surface at a point, or a *shear stress* acting *tangentially, in plane,* at the point. Up to now, the choice of planes upon which these stress components act, their orientation within a solid, was dictated by the geometry of the solid and the nature of the loading. We have said nothing about how these stress components might change if we looked at a set of planes of another orientation at the point. And up to now, we have said little about how these normal and shear stresses might vary with position throughout a solid.[1]

Now we consider a more general situation, namely an arbitrarily shaped solid which may be subjected to all sorts of externally applied loads - distributed or concentrated forces and moments. We are going to lift our gaze up from the world of crude structural elements such as truss bars in tension, shafts in torsion, or beams in bending to view these "solids" from a more abstract perspective. They all become special cases of a more general stuff we call a solid continuum.

Likewise, we develop a more general and more abstract representation of internal forces, moving beyond the notions of shear force, internal torque, uni-axial tension or compression and internal bending moment. Indeed, we have already done so in our representation of the internal force in a truss element as a normal stress, in our representation of torque in a thin-walled, circular shaft as the resultant of a uniformly distributed shear stress, in our representation of internal forces in a cylinder under internal pressure as a hoop stress (and as an axial stress). We want to develop our vocabulary and vision in order to speak intelligently about stress in its most general form.

We address two questions:

- How do the normal and shear components of stress acting on a plane at a given point change as we change the orientation of the plane at the point.

- How might stresses vary throughout a continuum;

The first bullet concerns the *transformation of components of stress at a point;* the second introduces the notion of *stress field.* We take them in turn.

---

1. The beam is the one exception. There we explored how different normal stress distributions over a rectangular cross-section could be equivalent to a bending moment and zero resultant force.

4.1   Stress: The Creature and its Components

We first address what we need to know to fully define "stress at a point" in a solid continuum. We will see that the stress at a point in a solid continuum is defined by its scalar components. Just as a vector quantity, say the velocity of a projectile, is defined by its *three* scalar components, we will see that the stress at a point in a solid continuum is defined by its *nine* scalar components.

Now you are probably quite familiar with vector quantities - quantities that have three scalar components. But you probably have not encountered a quantity like stress that requires more than three scalar values to fix its value at a point. This is a new kind of animal in our menagerie of variables; think of it as a new species, a new creature in our zoo. But don't let the number nine trouble you. It will lead to some algebraic complexity, compared to what we know how to do with vectors, but we will find that stress, a second order tensor, behaves as well as any vector we are familiar with.

The figure below is meant to illustrate the more general, indeed, the most general state of stress at a point. It requires some explanation:

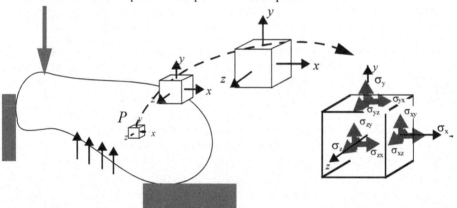

The odd looking structural element, fixed to the ground at bottom and to the left, and carrying what appears to be a uniformly distributed load over a portion of its bottom and a concentrated load on its top, is meant to symbolize an arbitrarily loaded, arbitrarily constrained, arbitrarily shaped solid continuum. It could be a beam, a truss, a thin-walled cylinder though it looks more like a potato — which too is a solid continuum. At any arbitrarily chosen point inside this object we can ask about the value of the stress at the point, say the point *P*. But what do we mean by "value"; value of what at that point?

Think about the same question applied to a vector quantity: What do we mean when we say we know the value of the velocity of a projectile at a point in its trajectory? We mean we know its magnitude - its "speed" a scalar - and its direction. Direction is fully specified if we know two more scalar quantities, e.g., the direction the vector makes with respect to the axes as measured by the cosine of the angles it makes with each axis. More simply, we have fully defined the velocity at a point if we specify its three scalar components with respect to some reference coordinate frame - say its x, y and z components.

Now how do we know this fully defines the vector quantity? We take as our criterion that anyone, anyone in the world (of mathematical physicists and engineers), would agree that they have in hand the same thing, no matter what coordinate frame they favor, no matter how they viewed the motion of the projectile. (We do insist that they are not displacing or rotating relative to one another, i.e., they all reside in the same inertial frame). This is assured if, after transforming the scalar components defined with respect to one reference frame to another, we obtain values for the components any other observer sees.

It is then the equations which transform the values of the components of the vector from one frame to another which define what a vector is. This is like defining a thing by what it does, e.g., "you are what you eat", a behaviorists perspective - which is really all that matters in mathematical physics and in engineering.

> Reid: Hey Katie: what do you think of all this talk about components? Isn't he going off the deep end here?
>
> Katie: What do you mean, "...off the deep end"?
>
> Reid: I mean why don't we stick with the stuff we were doing about beams and trusses? I mean that is the useful stuff. This general, abstract continuum business does nothing for me.
>
> Katie: There must be a reason, Reid, why he is doing this. And besides, I think it is interesting; I mean have you ever thought about what makes a vector a vector?
>
> Reid: I know what a vector is; I know about force and velocity; I know they have direction as well as a magnitude. So big deal. He is maybe just trying to snow us with all this talk about transformations.
>
> Katie: But the point is what makes force and velocity the same thing?
>
> Reid: They're not the same thing!
>
> Katie: He is saying they are - at a more abstract, general level. Like....like robins and bluebirds are both birds.
>
> Reid: And so stress is like tigers, is that it?
>
> Katie: Yeah, yeah - he said a new animal in the zoo.
>
> Reid: Huh... pass me the peanuts, will you?

We envision the components of stress as coming in sets of three: One set acts upon what we call an *x plane,* another upon a *y plane*, a third set upon a *z plane.* Which plane is which is defined by its *normal*: An *x plane has its normal in the x direction*, etc. Each set includes three scalar components, one *normal stress component* acting perpendicular to its reference plane, with its direction along one coordinate axis, and two *shear stress components* acting *in plane* in the direction of the other two coordinate axes.

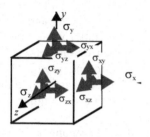

That's a grand total of *nine stress components to define the stress at a point.* To fully define the stress field throughout a continuum you need to specify how these components vary from point to point. Fortunately, equilibrium requirements applied to a differential element of the continuum, what we will call a "micro-equilibrium" consideration, will reduce the number of independent stress components at a point from nine to six. We will find that the shear stress component σ_{xy} acting on the x face must equal its neighbor around the corner σ_{yx} acting on the y face and that σ_{zy} = σ_{yz} and σ_{xz} = σ_{zx} accordingly.

Fortunately too, in most of the engineering structures you will encounter, diagnose or design, only two or three of these now six components will matter, that is, will be *significant*. Often variations of the stress components in one, or more, of the three coordinate directions may be uniform. But perhaps the most important simplification is a simplification in modeling, made at the outset of our encounter. One particularly useful model, applicable to many structural elements is called *Plane Stress* and, as you might infer from the label alone, it restricts our attention to variations of stress in two dimensions.

Plane Stress

If we assume our continuum has the form of a thin plate of uniform thickness but of arbitrary closed contour in the x-y plane, our previous arbitrarily loaded, arbitrarily constrained continuum (we don't show these again) takes the planar form below.

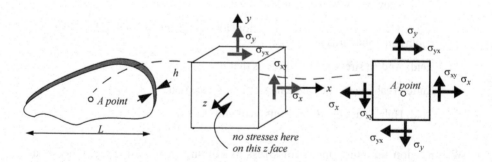

Because the plate is thin in the z direction, ($h/L \ll 1$) we will assume that variations of the stress components with z is uniform or, in other words, our stress components will be at most functions of x and y. We also take it that the z *boundary planes are unloaded, stress-free.* These two assumptions together imply that the set of three "z" stress components that act upon *any* arbitrarily located z plane within the interior must also vanish. We will also take advantage of the micro-equilibrium consequences, yet to be explored but noted previously, and set σ_{yz} and σ_{xz} to zero. Our state of stress at a point is then as it is shown on the exploded view of the point - the block in the middle of the figure - and again from the point of view of looking normal to a z plane at the far right. This special model is called **Plane Stress**.

A Word about Sign Convention:

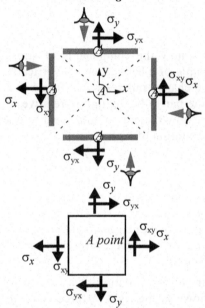

The figure at the far right seems to include more stress components than necessary; after all, if, in modeling, we eliminate the stress components acting on a *z* face and σ_{yz} and σ_{xz} as well, that should leave, at most, four components acting on the *x* and *y* faces. Yet there appear to be eight in the figure. No, there are only at most four components; we must learn to read the figure.

To do so, we make use of another sketch of *stress at a point*, the point *A*. The figure at the top is meant to indicate that we are looking at four faces or planes simultaneously. When we look at the *x face* from the right we are looking at the stress components on *a positive x face* — it has its *outward normal in the positive x direction* — and a *positive normal stress*, by convention, is *directed in the positive x direction*. A *positive shear stress component*, acting in plane, also acts, by convention, *in a positive coordinate direction* - in this case the positive y direction.

On the positive y face, we follow the same convention; *a positive* σ_y *acts on a positive y face in the positive y coordinate direction; a positive* σ_{yx} *acts on a positive y face in the positive x coordinate direction.*

We emphasize that we are looking at a point, point A, in these figures. More precisely we are looking at two mutually perpendicular planes intersecting at the point and from two vantage points in each case. We draw these two views of the two planes as four planes in order to more clearly illustrate our sign convention. But you ought to imagine the square having zero height and width: the σ_x acting to the *left*, in the *negative x direction*, upon the *negative x face* at the left, with its *outward normal pointing in the negative x direction* is *a positive component* at the point, the **equal and opposite reaction to the** σ_x acting to the right, in the *positive x direction*, upon the *positive x face* at the right, with its *outward normal pointing in the positive x direction*. Both are positive as shown; **both are the same quan-**

tity. So too the shear stress component σ_{xy} shown acting down, *in the negative y direction,* on the *negative x face* is the equal and opposite internal reaction to σ_{xy} shown acting up, *in the positive y direction,* on the *positive x face.*

A general statement of our sign convention, which holds for all nine components of stress, even in 3D, is as follows:

A positive component of stress acts on a positive face in a positive coordinate direction *or* on a negative face in a negative coordinate direction.

Transformation of Components of Stress

Before constructing the equations which fix how the components of stress transform in general, we consider a simple example of a bar suspended vertically and illustrate how components change when we change our reference frame at a point. In this example, we take the weight of the bar to be negligible relative to the weight suspended at its free end and explore how the normal and shear stress components at a point vary as we change the orientation of a plane.

Exercise 4.1

The solid column of rectangular cross section measuring a × b supports a weight W. Show that both a normal stress and a shear stress must act on any inclined interior face. Determine their respective values assuming that both are uniformly distributed over the area of the inclined face. Express your estimates in terms of the ratio (W/ab) and the angle φ.

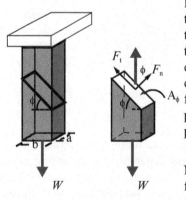

For equilibrium of the isolation of a section of the column shown at the right, a force equal to the suspended weight (we neglect the weight of the column itself) must act upward. We show an equivalent force system — or, if you like, its components consisting of two perpendicular forces, one directed normal to the inclined plane, the other with its line of action in the plane inclined at the angle φ. We have

$$F_n = W \cdot \cos\phi \qquad and \qquad F_t = W \cdot \sin\phi$$

Now if we assume these are distributed uniformly over the section, we can construct an estimate of the normal stress and the shear stress acting on the inclined face. But first we must establish the area of the inclined face $A\phi$. From the geometry of the figure we see that the length of the inclined plane is $b/\cos\phi$ so the area is

$$A_\phi = (ab)/(\cos\phi)$$

With this we write the normal and shear stress components as

$$\sigma_n = F_n / A_\phi = \left(\frac{W}{ab}\right) \cdot \cos\phi^2 \qquad and \qquad \sigma_t = F_t / A_\phi = \left(\frac{W}{ab}\right) \cdot \cos\phi\sin\phi$$

These results clearly illustrate how the values for the normal and shear stress components of a force distributed over a plane inside of an object depends upon how you look at the point inside the object in the sense that **the values of the shear and normal stresses at a point within a continuum depend upon the orientation of the plane you have chosen to view.**

Why would anyone want to look at some arbitrarily oriented plane in an object, seeking the normal and shear stresses acting on the plane? Why do we ask you to learn how to figure out what the stress components on such a plane might be?

The answer goes as follows: One of our main concerns as a designer of structures is failure —fracture or excessive deformation of what we propose be built and fabricated. Now many kinds of failures initiate at a local, microscopic level. A minute imperfection at a point in a beam where the local stress is very high initiates fracture or plastic deformation, for example. Our quest then is to figure out where, at what points in a structural element, the normal and shear stress components achieve their maximum values. But we have just seen how these values depend upon the way we look at a point, that is, upon the orientation of the plane we choose to inspect. To ensure we have found the maximum normal stress at a point for example, we would then have to inspect every possible orientation of a plane passing through the point.[2]

This seems a formidable task. But before taking it on, we pose a prior question:

Exercise 4.2

What do you need to know in order to determine the normal and shear stress components acting upon an arbitrarily oriented plane at a point in a fully three dimensional object?

The answer is what we might anticipate from our original definition of six stress components for if we know these six scalar quantities[3], the three normal stress components σ_x, σ_y, and σ_z, and the three shear stress components σ_{xy}, σ_{yz}, and σ_{xz}, then we can find the normal and shear stress components acting upon an arbitrarily oriented plane at the point. That is the answer to our need to know question.

To show this, we derive a set of equations that will enable you to do this. But note: we take the six stress components relative to the three orthogonal, let's call

---

2. Much as we have done in the preceding exercise. Our analysis shows that the maximum normal stress acts on the horizontal plane, defined by $\phi = 0$. The maximum shear stress, on the other hand acts on a plane oriented at 45° to the horizontal. The factor $\cos\phi \sin\phi$ has a maximum at $\phi = 45°$.

3. We take advantage of moment equilibrium and take $\sigma_{yx} = \sigma_{xy}$, $\sigma_{zx} = \sigma_{xz}$, and $\sigma_{zy} = \sigma_{yz}$.

them, x,y,z planes as given, as known quantities.   Furthermore, again we restrict our attention to two dimensions - the case of Plane Stress. That is we say that the components of stress acting on one of the planes at the point - we take the z planes - are zero. This is a good approximation for certain objects — those which are *thin* in the z direction relative to structural element's dimensions in the x-y plane. It also makes our derivation a bit less tedious, though there is nothing conceptual complex about carrying it through for three dimensions, once we have it for two.

In two dimensions we can draw a simpler picture of *the state of stress at a point*. We are not talking differential element here but of stress at a point. The figure below shows an arbitrarily oriented plane, defined by its normal, the *x'* axis, inclined at an angle ϕ to the horizontal. In this two dimensional state of stress we

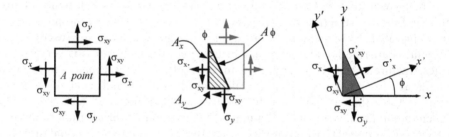

have but three scalar components to specify to fully define the state of stress at a point: σ_x, σ_y and $\sigma_{yx} = \sigma_{xy}$. Knowing these three numbers, we can determine the normal and shear stress components acting on any plane defined by the orientation ϕ as follows.

Consider equilibrium of the shaded wedge shown. Here we let $A\phi$ designate the area of the inclined face at a point, A_x and A_y the areas of the x face with its outward normal pointing in the -x direction and of the y face with its outward normal pointing in the -y direction respectively. In this we take a unit depth into the paper. We have

$$A_x = A_\phi \cdot \cos\phi \qquad and \qquad A_y = A_\phi \cdot \sin\phi$$

That takes care of the relative areas. Now for *force* equilibrium, in the x and y directions we must have:

$$-\sigma_x \cdot A_x - \sigma_{xy} \cdot A_y + (\sigma'_x \cdot \cos\phi - \sigma'_{xy} \cdot \sin\phi) \cdot A_\phi = 0$$

$$and$$

$$-\sigma_{xy} \cdot A_x - \sigma_y \cdot A_y + (\sigma'_x \cdot \sin\phi + \sigma'_{xy} \cdot \cos\phi) \cdot A_\phi = 0$$

If we multiply the first by $\cos\phi$, the second by $\sin\phi$ and add the two we can eliminate σ'_{xy}. We obtain

$$\sigma'_x A_\phi - \sigma_x \cos\phi A_x - \sigma_{xy}\cos\phi A_y - \sigma_{xy}\sin\phi A_x - \sigma_y \sin\phi A_y = 0$$

which, upon expressing the areas of the *x,y* faces in terms of the area of the inclined face, can be written (noting $A\phi$ becomes a common factor).

$$\sigma'_x = \sigma_x cos\phi^2 + \sigma_y sin\phi^2 + 2\sigma_{xy} sin\phi cos\phi$$

In much the same way, multiplying the first equilibrium equation by $sin\phi$, the second by $cos\phi$ but subtracting rather than adding you will obtain eventually

$$\sigma'_{xy} = (\sigma_y - \sigma_x) sin\phi cos\phi + \sigma_{xy}(cos\phi^2 - sin\phi^2)$$

We deduce the normal stress component acting on the y' face of this rotated frame by replacing ϕ in our equation for σ'_x by $\phi + \pi/2$. We obtain in this way:

$$\sigma'_y = \sigma_y cos\phi^2 + \sigma_x sin\phi^2 - 2\sigma_{xy} sin\phi cos\phi$$

The three transformation equations for the three components of stress at a point can be expressed, using the double angle formula for the cosine and the sine, as

$$\sigma'_x = \left[\frac{(\sigma_x + \sigma_y)}{2}\right] + \left[\frac{(\sigma_x - \sigma_y)}{2}\right] \cdot cos2\phi + \sigma_{xy} sin2\phi$$

$$\sigma'_y = \left[\frac{(\sigma_x + \sigma_y)}{2}\right] - \left[\frac{(\sigma_x - \sigma_y)}{2}\right] \cdot cos2\phi - \sigma_{xy} sin2\phi$$

$$\sigma'_{xy} = -\left[\frac{(\sigma_x - \sigma_y)}{2}\right] \cdot sin2\phi + \sigma_{xy} cos2\phi$$

Here we have the equations to do what we said we could do. Think of the set as a machine: You input the three components of stress at a point defined relative to an x-y coordinate frame, then give me the angle ϕ, and I will crank out -- not only the normal and shear stress components acting on the face with its outward normal inclined at the angle ϕ with respect to the x axis, but the normal stress on the y' face as well. In fact I could draw a square tilted at an angle ϕ to the horizontal and show the stress components σ'_x, σ'_y and σ'_{xy} acting on the x' and y' faces.

To show the utility of these relationships consider the following scenario:

Exercise 4.3

An solid circular cylinder made of some brittle material is subject to pure torsion —a torque M_t. If we assume that a shear stress $\tau(r)$ acts within the cylinder, distributed over any cross section, varying with r according to

$$\tau(r) = c \cdot r^n$$

where n is a positive integer, then the maximum value of τ, will occur at the outer radius of the shaft.

But is this the maximum value? That is, while certainly r'' is maximum at the outermost radius, $r=R$, it may very well be that the maximum shear stress acts on some other plane at that point in the cylinder.

Show that the maximum shear stress is indeed that which acts on a plane normal to the axis of the cylinder at a point on the surface of the shaft.

Show too, that the maximum normal stress in the cylinder acts

- at a point on the surface of the cylinder

- on a plane whose normal is inclined 45° to the x axis and its value is
$$\sigma'_x\big|_{max} = \tau(R)$$

We put to use our machinery for computing the stress components acting upon an arbitrarily oriented plane at a point. Our initial set of stress components for this particular state of stress is

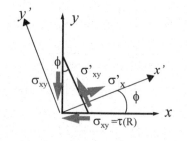

$$\sigma_x = 0$$

$$\sigma_y = 0$$

and

$$\sigma_{xy} = \tau(R)$$

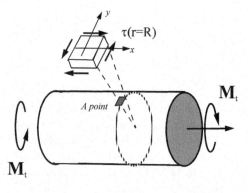

defined relative to the x-y coordinate frame shown top right. Our equations defining the transformation of components of stress at the point take the simpler form

$$\sigma'_x = \tau \cdot sin2\phi$$

$$\sigma'_y = -\tau \cdot sin2\phi$$

$$\sigma'_{xy} = \tau \cdot cos2\phi$$

To find the maximum value for the shear stress component with respect to the plane defined by ϕ, we set the derivative of σ'_{xy} to zero. Since there are no "boundaries" on ϕ to worry about, this ought to suffice.

So, for a maximum, we must have

$$\frac{d\sigma'_{xy}}{d\phi} = -2\tau \cdot sin2\phi = 0$$

Now there are many values of ϕ which satisfy this requirement, $\phi=0$, $\phi=\pi/2$, ... But all of these roots just give the orientation of our initial two mutually perpendicular, x-y planes. Hence the maximum shear stress within the shaft is just τ at r=R.

To find the extreme, including maximum, values for the normal stress, σ'_x we proceed in much the same way; differentiating our expression above for σ'_x with respect to ϕ yields

$$\frac{d\sigma'_x}{d\phi} = 2\tau \cdot \cos 2\phi = 0 \qquad \text{(EQ 1)}$$

Again there is a string of values of ϕ, each of which satisfies this requirement.

We have $2\phi = \pi/2$, $3\pi/2$ .... or $\phi = \pi/4$, $3\pi/4$

At $\phi = \pi/4$ (= 45°), the value of the normal stress is $\sigma'_x = + \tau \sin 2\phi = \tau$. So the maximum normal stress acting at the point on the surface is equal in magnitude to the maximum shear stress component.

Note too that our transformation relations say that the normal stress component acting on the y' plane, with $\phi=\pi/4$ is negative and equal in magnitude to τ. Finally we find that the shear stress acting on the x'-y' planes is zero! We illustrate the state of stress at the point relative to the x'-y' planes below right.

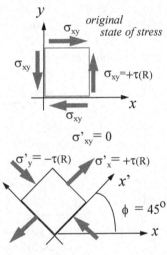

Backing out of the woods in order to see the trees, we claim that if our cylinder is made of a brittle material, it will fracture across the plane upon which the maximum tensile stress acts. If you go now and take a piece of chalk and subject it to a torque until it breaks, you should see a fracture plane in the form of a helical surface inclined at 45 degrees to the axis of the cylinder. Check it out.

Of course it's not enough to know the orientation of the fracture plane when designing brittle shafts to carry torsion. We need to know the *magnitude of the torque* which will cause fracture. In other words we need to know how the shear stress does *in fact* vary throughout the cylinder.

This remains an unanswered question. So too for the beam: How do the normal stress (and shear stress) components vary over a cross section of the beam? In a subsequent section, we explore how far we can go with equilibrium consideration in responding. But, in the end, we will find that the problem remains statically indeterminate; we will have to go beyond the concept of stress and consider the deformation and displacement of points in the continuum. But first, a special technique for doing the transformation of components of stress at a point. "Mohr's

Circle" is a graphical technique which, while offering no new information, does provide a different and useful perspective on our subject.

Studying Mohr's Circle is customarily the final act in this first stage of indoctrination into concept of stress. Your uninitiated colleagues may be able to master the idea of a truss member in tension or compression, a beam in bending, a shaft in torsion using their common sense knowledge of the world around them, but Mohr's Circle will appear as a complete mystery, an unfathomable ritual of signs, circles, and greek symbols. Although it does not tell us anything new, over and above all that we have done up to this point in the chapter, once you've mastered the technique it will set you apart from the crowd and shape your very well being. It may also provide you with a useful aid to understanding the transformation of stress and strain at a point on occasion.

Mohr's Circle

Our working up of the transformation relations for stress and our exploration of their implications for determining extreme values has required considerable mathematical manipulation. We turn now to a graphical rendering of these relationships. I will set out the rules for constructing the circle for a particular state of stress, show how to read the pattern, then comment about its legitimacy. I first repeat the transformation equations for a two-dimensional state of stress.

$$\sigma'_x = \left[\frac{(\sigma_x + \sigma_y)}{2}\right] + \left[\frac{(\sigma_x - \sigma_y)}{2}\right] \cdot \cos 2\phi + \sigma_{xy} \sin 2\phi$$

$$\sigma'_y = \left[\frac{(\sigma_x + \sigma_y)}{2}\right] - \left[\frac{(\sigma_x - \sigma_y)}{2}\right] \cdot \cos 2\phi - \sigma_{xy} \sin 2\phi$$

$$\sigma'_{xy} = -\left[\frac{(\sigma_x - \sigma_y)}{2}\right] \cdot \sin 2\phi + \sigma_{xy} \cos 2\phi$$

To construct Mohr's Circle, given the state of stress $\sigma_x = 7$, $\sigma_{xy} = 4$, *and* $\sigma_y = 1$ we proceed as follows: Note that I have dropped all pretense of reality in this choice of values for the components of stress. As we shall see, it is their relative magnitudes that is important to this geometric construction. Everything will scale by any common factor you please to apply. You could think of these as $\sigma_x = 7 \times 10^3$ KN/m²...etc., if you like.

- Lay out a horizontal axis and label it σ positive to the right.

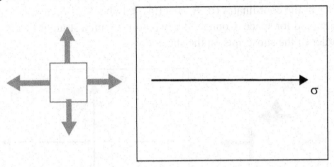

- Lay out an axis perpendicular to the above and label it σ_{xy} **positive down** and σ_{yx} **positive up**[4].

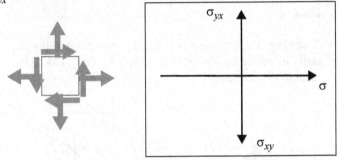

- Plot a point associated with the stress components acting on an x face at the coordinates $(\sigma_x, \sigma_{xy})=(7, 4_{down})$. Label it x_{face}, or x if you are cramped for space.

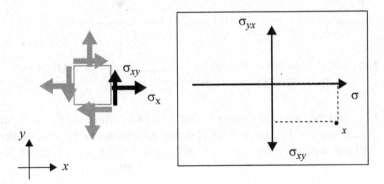

---

4. WARNING: Different authors and engineers use different conventions in constructing the Mohr's circle.

- Plot a second point associated with the stress components acting on an y face at the coordinates $(\sigma_y, \sigma_{yx}) = (1, 4_{up})$. Label it y_{face}, or y if you are cramped for space. Connect the two points with a straight line. Note the order of the subscripts on the shear stress.

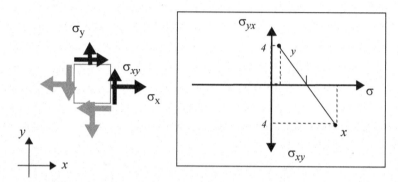

- Chanting "similar triangles", note that the center of the line must necessarily lie on the horizontal, σ axis since $\sigma_{xy} = \sigma_{yx}$, 4=4. Draw a circle with the line as a diagonal.

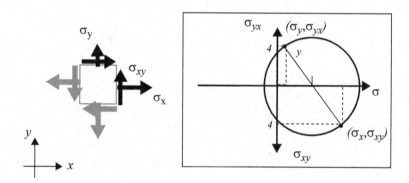

- Note that the radius of this circle is $R_{Mohrs} = \sqrt{(\sigma_{xy})^2 + [(\sigma_x - \sigma_y)/2]^2}$ which for the numbers we are using is just $R_{Mohr's\ C} = 5$, and its center lies at $(\sigma_x + \sigma_y)/2 = 4$.
- To find the stress components acting on a plane whose normal is inclined at an angle of ϕ degrees, positive counterclockwise, to the x axis **in the physical plane**, rotate the diagonal 2ϕ **in the Mohr's Circle plane**. We illustrate this for $\phi = 40^o$. Note that the shear stress on the new x' face is

negative according to the convention we have chosen for our Mohr's Circle.[5]

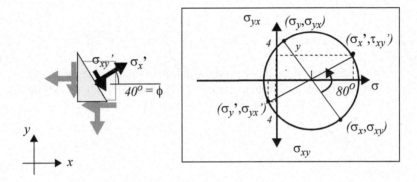

- The stress components acting on the y' face, at $\phi + \pi/2 = 130^o$ around in the **physical plane** are $2\phi + \pi = 240^o$ around in the **Mohr's Circle plane**, just 2ϕ around from the y face in the Mohr's Circle plane.

We establish the legitimacy of this graphical representation of the transformation equations for stress making the following observations:

- The extreme values of the normal stress lie at the two intersections of the circle with the σ axis. The angle of rotation from the x_{face} to the principal plane I on the Mohr's Circle is related to the stress components by the equation previously derived:

$$\tan 2\phi = 2\sigma_{xy}/(\sigma_x - \sigma_y).$$

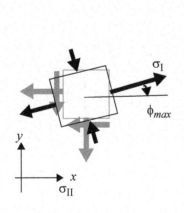
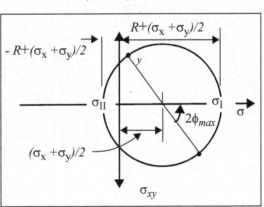

---

5. WARNING, again: Other texts use other conventions.

- Note that on the principal planes the shear stress vanishes.

- The values of the two principal stresses can be written in terms of the radius of the circle.

$$\sigma_{I,II} = [(\sigma_x + \sigma_y)/2] \pm \sqrt{(\sigma_{xy})^2 + [(\sigma_x - \sigma_y)/2]^2}$$

- The orientation of the planes upon which an extreme value for the shear stress acts is obtained from a rotation of 90° around from the σ axis on the Mohr's Circle. The corresponding rotation in the physical plane is 45°.

- The sum $\sigma_x + \sigma_y$ is an invariant of the transformation. The center of the Mohr's Circle does not move. This result too can be obtained from the equations derived simply by adding the expression for σ_x' to that obtained for σ_y'.

- So too the radius of the Mohr's Circle is an invariant. This takes a little more effort to prove.

Enough. Now onto the second topic of the chapter - the variation of stress components as we move throughout the continuum. This is prerequisite if we seek to find extreme values of stress.

4.2   The Variation of Stress (Components) in a Continuum

To begin, we re-examine the case of a bar suspended vertically but now consider the state of stress at each and every point in the continuum engendered by its own weight. (Note, I have changed the orientation of the reference axes). We will

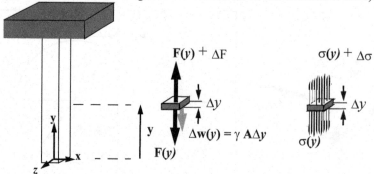

construct a differential equation which governs how the axial stress varies as we move up and down the bar. We will solve this differential equation, not forgetting to apply an appropriate *boundary condition* and determine the axial stress field.

We see that for equilibrium of the differential element of the bar, of planar cross-sectional area A and of weight density γ, we have

$$F + \Delta F - \gamma \cdot A \cdot \Delta y - F = 0$$

If we assume the tensile force is uniformly distributed over the cross-sectional area, and dividing by the area (which does not change with the independent spatial coordinate y) we can write

$$\sigma + \Delta \sigma - \gamma \cdot \Delta y - \sigma = 0 \qquad where \qquad \sigma \equiv F/A$$

Chanting "...going to the limit, letting Δy go to zero", we obtain a differential equation fixing how $\sigma(y)$, a function of y, varies throughout our continuum, namely

$$\frac{d\sigma}{dy} - \gamma = 0$$

We solve this ordinary differential equation easily, integrating once and obtain

$$\sigma(y) = \gamma \cdot y + Constant$$

The *Constant* is fixed by a prescribed condition at some y surface; If the end of the bar is *stress free*, we indicate this writing

$$at\ y = 0 \qquad \sigma = 0$$

$$so$$

$$\sigma(y) = \gamma \cdot y$$

If, on another occasion, a weight of magnitude P_0 is suspended from the free end, we would have

$$at\ y = 0 \qquad \sigma = P_0/A$$

$$and$$

$$\sigma(y) = \gamma \cdot y + P_0/A$$

Here then are two stress fields for two different loading conditions[6]. Each stress field describes how the normal stress $\sigma(x,y,z)$ varies throughout the continuum at every point in the continuum. I show the stress as a function of x and z as well as y to emphasize that we can evaluate its value at *every point* in the continuum, although it only varies with y. That the stress does not vary with x and z was implied when we stipulated or assumed that the internal force, F, acting upon any y plane was uniformly distributed over that plane. This example is a special case in another way; not only is it one-dimensional in its dependence upon spatial position, but it is the simplest example of stress at a point in that it is described fully by a single component of stress, the normal stress acting on a plane perpendicular to the y axis.

---

6. A third loading condition is obtained by setting the weight density γ to zero; our bar then is assumed weightless relative to the end-load P_0.

Stress Fields & "Micro" Equilibrium

In our analysis of how the normal stress varied throughout the vertically suspended bar, we considered a differential element of the bar and constructed a differential equation which described how the normal stress component varied in one
direction, in one spatial dimension.   We can call this picture of equilibrium
"micro" in nature and distinguish it from the "macro" equilibrium considerations
of the last chapter. There we isolated large chunks of structure e.g., when we cut
through the beam to see how the shear force and bending moment varied with distance along the beam.

Now we look with finer resolution and attempt to determine how the normal
and shear stress components vary at the micro level throughout the beam.   The
question may be put this way: Knowing the shear force and bending moment at
any section along the beam, how do the normal and shear stress components vary
over the section?

To proceed, we make some appropriate assumptions about the nature of the
beam and build upon the conjectures we made in the last chapter about how the
stress components might vary.

We model the end-loaded cantilever with a relatively thin rectangular cross-section as a plane stress problem. In this, b is the "thin" dimension, i.e., $b/L < 1$.

If we assume a normal stress distribution *over an x face* is proportional to some odd power of y, as we did in Chapter 3, our state of stress at a point might look like that shown in figure (c). In this, σ_x would have the form

$$\sigma_x(x, y) = C(n, b, h) \cdot W(L - x)y^n$$

where C(n,b,h) is a constant which depends upon the cross-sectional dimensions of the beam and the odd exponent n. The factor W(L-x) is the magnitude of the internal bending moment at the location x measured from the root. See figure (b).

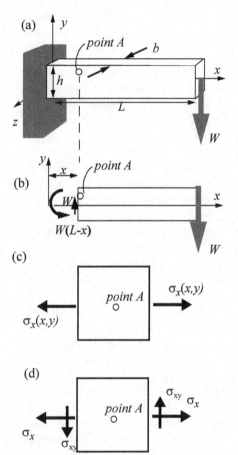

But this is only one component of our stress field. What are the other components of stress at point A?

Our plane stress model allows us to claim that the three *z face* components are zero and if we take σ_{yz} and σ_{xz} to be zero, that still leaves σ_{xy}, and σ_y in addition to σ_x.

To continue our estimation process, we make the most of what we already know: For example, we know that a shear force of magnitude W acts at any *x* section. For the end-loaded cantilever, neglecting the weight of the beam itself, it does not vary with x. We might assume that the shear force is uniformly distributed over the cross-section and set $\sigma_{xy} = -W/(bh)$. Our stress at point A would then look like figure (d).

We could, of course, posit other shear stress distributions at any x station, e.g., some function like $\sigma_{xy} = Constant \cdot y^m$ where m is an integer and the constant is determined from the requirement that the resultant force due to this shear stress distribution over the cross section must be W.

The component σ_y - how it varies with x and y - remains a complete unknown. We will argue that it is small, relative to the normal and shear stress components, moved by the observation that the normal stress on the top and the bottom surfaces of the beam is zero; we say the top and bottom surfaces are "stress free". Continuing, if σ_y vanishes there at the bounary, then it probably will not grow to be significant in the interior. This indeed can be shown to be the case if h/L $\ll$ 1, as it is for a beam. So we estimate σ_y=0.

But there is something more we can do. We can look at equilibrium of a differerential element within the beam and, as we did in the case of a bar hung vertically, construct a differential equation whose solution (supplemented with suitable boundary conditions) defines how the normal and shear stress components vary throughout the plane, with x and y. Actually we construct more than a single differential equation: We obtain two, coupled, first-order, partial differential equations for the normal and shear stress components.

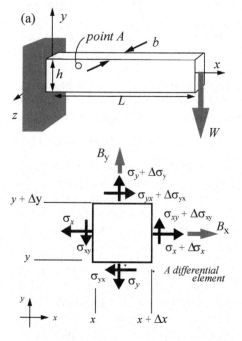

Think now, of a differential element in 2D at any point within the cantilever beam: We show such on the right. Note now we are no longer focused on two intersecting, perpendicular planes at a point but on a differential element of the continuum. Now we see that the stress components may very well be different on the two *x* faces and on the two *y* faces.

We allow the x face components, and those on the two y faces to change as we move from *x* to *x*+Δx (holding y constant) and from *y* to *y*+Δy (holding *x* constant).

We show two other arrows on the figure, B_x and B_y. These are meant to represent the x and y components of what is called a *body force*. A body force is any externally applied force acting on each element of volume of the continuum. It is thus a force per unit volume. For example, if we need consider the weight of the beam, B_y would be just

$$B_y = -\gamma \qquad \text{where} \qquad \gamma = \text{the weight density}$$

where the negative sign is necessary because we take a positive component of the body force vector to be in a positive coordinate direction.

B_x would be taken as zero.

We now consider force and moment equilibrium for this differential element, our micro isolation. We sum forces in the *x* direction which will include the shear stress component σ_{yx}, acting on the *y* face in the *x* direction as well as the normal stress component σ_x acting on the x faces. But note that these components are not forces; to figure their contribution to the equilibrium requirement, we must factor in the areas upon which they act.

I present just the results of the limiting process which, we note, since all components may be functions of both x and y, brings partial derivatives into the picture.

$$\sum \text{Forces in the x direction} \rightarrow \qquad \frac{\partial \sigma_x}{\partial x} + \frac{\partial \sigma_{yx}}{\partial y} + B_x = 0$$

$$\sum \text{Forces in the y direction} \rightarrow \qquad \frac{\partial \sigma_{xy}}{\partial x} + \frac{\partial \sigma_y}{\partial y} + B_y = 0$$

$$\sum \text{Moments about the center of the element} \rightarrow \qquad \sigma_{yx} = \sigma_{xy}$$

For example, the change in the stress component $\Delta\sigma_x$ may be written

$$\Delta\sigma_x = \frac{\partial \sigma_x}{\partial x} \cdot \Delta x$$

and the force due to this "unbalanced" component in the x direction is

$$\frac{\partial \sigma_x}{\partial x} \cdot \Delta x \cdot (\Delta y \Delta z)$$

where the product, $\Delta y \, \Delta z$, is just the differential area of the *x face*.

The contribution of the body force (per unit volume) to the sum of force components in the x direction will be $B_x(\Delta x \Delta y \Delta z)$ where the product of deltas is just the differential volume of the element. We see that this product will be a common factor in all terms entering into the equations of force equilibrium in the x and y directions.

The last equation of moment equilibrium shows that, as we forecast, the shear stress component on the *y face* must equal the shear stress component acting on the *x* face. The differential changes in the shear stress components are of lower order and drop out of consideration in the limiting process, as we take Δx and Δy to zero.

We might now try to solve this system of differential equations for σ_{xy}, and σ_y and σ_x but, in fact, we are doomed from the start. Even with the simplification afforded by moment equilibrium we are left with two coupled, linear, first-order partial differential equations for these three unknowns. The problem is statically indeterminate so we are not going to be able to construct a unique solution to the equilibrium requirements.

To answer these questions we must go beyond the concepts and principles of static equilibrium. We have to consider the requirements of continuity of displacement and compatibility of deformation. This we do in the next chapter, looking first at simple indeterminate systems, then on to the indeterminate truss, the beam in bending and the torsion of shafts.

4.3 Problems

4.1 A fluid can be defined as a continuum which - unlike a solid body - is unable to support a shear stress and remain at rest. The state of stress at any point, within a fluid column for example, we label "hydrostatic"; the normal stress components are equal to the negative of the static pressure at the point and the shear stress components are all zero. $\sigma_x = \sigma_y = \sigma_z = -p$ and $\sigma_{xy} = \sigma_{xz} = \sigma_{yz} = 0$.

Using the two dimensional transformation relations (the existence of σ_z does not affect their validity) show that the shear stress on any arbitrarily oriented plane is zero and the normal stress is again -p.

4.2 Estimate the compressive stress at the base of the Washington Monument - the one on the Mall in Washington, DC.

4.3 A closed can of soda is under pressure equivalent to that in an automobile tire. We measured the wall thickness in lab on Thursday to be 0.0025 inches. Estimate the axial stress σ_a and the hoop stress σ_θ? acting at a point in this closed, thin cylindrical shell, away from the ends.

4.4 Estimate the "hoop stress" within an un-opened can of soda.

4.5 The stress at a point in the plane of a thin plate is shown. Only the shear stress component is not zero relative to the x-y axis. From equilibrium of a section cut at the angle ϕ, deduce expressions for the normal and shear stress components acting on the inclined face of area A. NB: stress is a force per unit area so the areas of the faces the stress components act upon must enter into your equilibrium considerations.

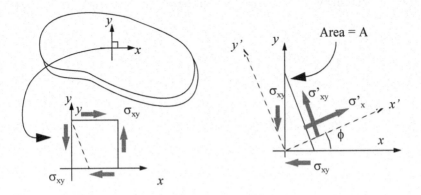

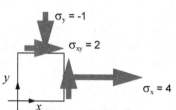

4.6 Given the components of stress relative to an x-y frame at a point in plane stress are:

$$\sigma_x = 4, \qquad \sigma_{xy} = 2 \qquad \sigma_y = -1$$

What are the components with respect to an axis system rotated 30° counterclockwise at the point?

4.7   Construct Mohr's circle for the state of stress of exercise **4.4**, above. Determine the "principle stresses" and the orientation of the planes upon which they act relative to the *xy* frame.

4.8   A thin walled glass tube of radius R = 1 inch, and wall thickness t= 0.010 inches, is closed at both ends and contains a fluid under pressure, p = 100 psi. A torque, M_t, of 300 inch-lbs, is applied about the axis of the tube.

Compute the stress components relative to a coordinate frame with its x axis in the direction of the tube's axis, its y axis circumferentially directed and tangent to the surface.

Determine the maximum tensile stress and the orientation of the plane upon which it acts.

4.9   *What if* we change our sign convention on stress components so that a normal, compressive stress is taken as a positive quantity (a tensile stress would then be negative). What becomes of the transformation relations? How would you alter the rules for constructing and using a Mohr's circle to find the stress components on an arbitrarily oriented plane?

What if you changed your sign convention on shear stress as well; how would things change?

5

Indeterminate Systems

The key to resolving our predicament, when faced with a statically indeterminate problem - one in which the equations of static equilibrium do not suffice to determine a unique solution - lies in opening up our field of view to consider the *displacements of points* in the structure and the *deformation of its members*. This introduces new variables, a new genera of flora and fauna, into our landscape; for the truss structure the species of *node displacements* and the related species of *uniaxial member strains* must be engaged. For the frame structure made up of beam elements, we must consider the *slope of the displacement* and the related *curvature of the beam* at any point along its length. For the shaft in torsion we must consider the rotation of one cross section relative to another.

Displacements you already know about from your basic course in physics – from the section on Kinematics within the chapter on Newtonian Mechanics. Displacement is a vector quantity, like force, like velocity; it has a magnitude and direction. In Kinematics, it tracks the movement of a physical point from some location at time t to its location at a subsequent time, say $t + \delta t$, where the term δt indicates a small time increment. Here, in this text, the displacement vector will, most often, represent the movement of a physical point of a structure from *its position in the undeformed state of the structure to its position in its deformed state*, from the structure's unloaded configuration to its configuration under load.

These displacements will generally be small relative to some nominal length of the structure. Note that previously, in applying the laws of static equilibrium, we made the tacit assumption that *displacements were so small we effectively took them as zero;* that is, we applied the laws of equilibrium to the *undeformed body.*[1] There is nothing inconsistent in what we did there with the tack we take now as long as we restrict our attention to small displacements. That is, our equilibrium equations taken with respect to the undeformed configuration remain valid even as we admit that the structure deforms.

Although *small* in this respect, the small displacement of one point *relative to* the small displacement of another point in the deformation of a structural member can engender large internal forces and stresses.

In the first part of this chapter, we do a series of exercises - some simple, others more complex - but all involving only one or two degrees of freedom; that is, they all concern systems whose deformed configuration is defined by but one or two displacements (and/or rotations). In the final part of this chapter, we consider

---

1. The one exception is the introductory exercise where we allowed the two bar linkage to "snap through"; in that case we wrote equilibrium with respect to the deformed configuration.

indeterminate truss structures - systems which may have many degrees of freedom. In subsequent chapters we go on to resolve the indeterminacy in our study of the shear stresses within a shaft in torsion and in our study of the normal and shear stresses within a beam in bending.

5.1   Resolving indeterminacy: Some Simple Systems.

If we admit displacement variables into our field of view, then we must necessarily learn how these are related to the forces which produce or are engendered by them. We must know how force relates to displacement. Force-displacement, or *constitutive* relations, are one of three sets of relations upon which the analysis of indeterminate systems is built. The requirements of *force and moment equilibrium* make up a second set; *compatibility of deformation* is the third.

A Word about Constitutive Relations

You are familiar with one such constitutive relationship, namely that between the force and displacement of a spring, usually a linear spring. $F = k \cdot \delta$ says that the force F varies linearly with the displacement δ. The spring constant (of proportionality) k has the *dimensions* of force/length. It's particular *units* might be pounds/inch, or Newtons/millimeter, or kilo-newtons/meter.

Your vision of a spring is probably that of a coil spring - like the kind you might encounter in a children's playground, supporting a small horse. Or you might picture the heavier springs that might have been part of the undercarriage of your grandfather's automobile. These are real-world examples of linear springs.

But there are other kinds that don't look like coils at all. A rubber band behaves like a spring; it, however, does not behave linearly once you stretch it an appreciable amount. Likewise an aluminum or steel rod when stretched behaves like a spring and in this case behaves linearly over a useful range - but you won't see the extension unless you have super-human eyesight.

For example, the picture at the left is meant to represent a rod, made of an aluminum alloy, drawn to full scale. It's length is L = 4 inches, its cross-sectional area A = 0.01 square inches. If we apply a force, F, to the free end as shown, the rod will stretch, the end will move downward just as a coil spring would. And, for small deflections, δ, if we took measurements in the lab, plotted force versus displacement, then measured the slope of what appears to be a straight line, we would have:

$$F = k \cdot \delta \qquad \text{where} \qquad k = 25{,}000 \text{ lb/inch}$$

This says that if we apply a force of 25,000 pounds, we will see an end displacement of 1.0 inch. You, however, will find that you can not do so.

The reason is that if you tried to apply a weight of this magnitude (more than 10 tons!) the rod would stretch more and more like a soft plastic. It would *yield* and fail. So there are limits to the loads we can apply to materials. That limit is a characteristic (and conventional) property of the material. For this particular aluminum alloy, the rod would fail at an axial stress of

σ_{yield} = 60,000 psi   or at a force level F = 600 pounds factoring in the

area of 0.01 square inches.

Note that at this load level, the end displacement, figured from the experimentally established stiffness relation, is δ = 0.024 inches (can you see that?) And thus the ratio δ/L is but 0.006. This is what we mean by small displacements. This is what we mean by linear behavior (only up to a point - in this case -the yield stress). This is the domain within which engineers design their structures (for the most part).

We take this as the way force is related to the displacement of individual structural elements in the exercises that follow[2].

Exercise 5.1

A massive stone block of weight W and uniform in cross section over its length L is supported at its ends and at its midpoint by three linear springs. Assuming the block a rigid body[3], construct expressions for the forces acting in the springs in terms of the weight of the block.

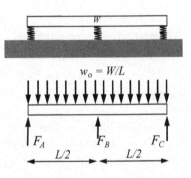

The figure shows the block resting on three linear springs. The weight per unit length we designate by $w_o = W/L$.

In the same figure, we show a free-body diagram.   The forces in the spring, taken as compressive, push up on the beam in reaction to the distributed load. Force equilibrium in the vertical direction gives:

---

2. We will have more to say about constitutive relations of a more general kind in a subsequent chapter.
3. The word *rigid* comes to the fore now that we consider the deformations and displacements of extended bodies. *Rigid* means that there is no, *absolutely no relative displacement of any two, arbitrarily chosen points in the body* when the body is loaded. Of course, this is all relative in another sense. There is always some relative displacement of points in each, every and all bodies; *a rigid body* is as much an abstraction as *a frictionless pin*. But in many problems, the relative displacements of points of some one body or subsystem may be assumed small relative to the relative displacements of another body. In this exercise we are claiming that the block of stone is rigid, the springs are not, i.e., they deform.

$$F_A + F_B + F_C - W = 0$$

While moment equilibrium, summing moments about the left end, A, taking counter-clockwise as positive, gives:

$$\sum_A M = F_B \cdot L/2 + F_C \cdot L - W \cdot L/2 = 0$$

The problem is *indeterminate*: Given the length L and the weight W, we have but *two* equations for the *three* unknown forces, the three compressive forces in the springs.

Now, indeterminacy *does not* mean we can not find a solution. What it *does* mean is that *we can not find* a *single, unambiguous, unique solution* for each of the three forces. That is what indeterminate means. We can find solutions - *too many* solutions; the problem is that we do not have sufficient information, e.g., enough equations, to fix which of the many solutions that satisfy equilibrium is the right one[4].

Indeterminate solution (to equilibrium alone) #1

For example, we might take $F_B = 0$, which in effect says we remove the spring support at the middle. Then for equilibrium we must subsequently have $F_A = F_C = W/2$ This is a solution to equilibrium.

Indeterminate solution (to equilibrium alone) #2

Alternatively, we might require that $F_A = F_C$; in effect adding a third equation to our system. With this we find from moment equilibrium that $F_A = F_C = W/3$ and so from force equilibrium $F_B = W/3$ This too is a solution.

Indeterminate solution (to equilibrium alone) #n, n=1,2,......

We can fabricate many different solutions in this way, an infinite number. For example, we might arbitrarily take $F_B = W/n$, where n = 1,2,....then from the two requirements for equilibrium find the other two spring forces. (Try it)!

Notice in the above that we have not said one word about the displacements of the rigid block nor a word about the springs, their stiffness, whether they are linear springs or non-linear springs. Now we do so. Now we really solve the indeterminate problem, setting three or four different scenarios, each defined by a different choice for the relative stiffness of the springs. In all cases, we will assume the springs are linear.

---

4. We say the equations of equilibrium are *necessary* but not *sufficient* to produce a solution.

Full Indeterminate solution, Scenario #1

In this first scenario, at the start, we assume also that they have equal stiffness.

We set

$$F_A = k \cdot \delta_A$$

$$F_B = k \cdot \delta_B \text{ where } \delta_A, \delta_B, \text{ and } \delta_C \text{ are the displacements of the springs,}$$

$$F_C = k \cdot \delta_C$$

taken positive downward since the spring forces were taken positive in compression[5]. The spring constants are all equal. These are the required *constitutive relations*.

Now *compatibility of deformation*: The question is, how are the three displacements related. Clearly they must be related; we can not choose them independently one from another, e.g., taking the displacements of the end springs as downward and the displacement of the midpoint as upwards. This could only be the case if the block had fractured into pieces. No, this can't be. We insist on compatibility of deformation.

Here we confront the same situation faced by Buridan's ass, that is, the situation to the left appears no different from the situation to the right so, "from symmetry" we claim there is no sufficient reason why the block should tip to the left or to the right. It must remain level[6].

In this case, the displacements are all equal.

$$\delta_A = \delta_B = \delta_C$$

This is our *compatibility* equation.

So, in this case, from the constitutive relations, the spring forces are all equal.
So, in this case,

$$F_A = F_B = F_C = W/3$$

Full Indeterminate solution, Scenario #2

In this second scenario, we assume the two springs at the end have the same stiffness, k, while the stiffness of the spring at mid-span is different. We set $k_B = \alpha k$ so our constitutive relations may be written

$$F_A = k \cdot \delta_A$$

$$F_B = \alpha k \cdot \delta_B \text{ where the non-dimensional parameter } \alpha \text{ can take on any posi-}$$

$$F_C = k \cdot \delta_C$$

tive value within the range 0 to very, very large.

---

5. We must be careful here; a positive force must correspond to a positive displacement.
6. Note that this would not be the case if the spring constants were chosen so as to destroy the symmetry, e.g., if $k_A > k_B > k_C$.

Notice again we have symmetry: There is still no reason why the block should tip to the left or to the right! So again, the three displacements must be equal.

$$\delta_A = \delta_B = \delta_C = \delta$$

The constitutive relations then say that the forces in the two springs at the end are equal, say $= F$ and that the force in the spring at mid span is αF.

With this, force equilibrium gives

$$F_A + F_B + F_C = W \qquad \text{i.e.,} \qquad (2 + \alpha) \cdot k \cdot \delta = W$$

So, in this scenario,

$$F_A = F_C = \frac{W}{(2 + \alpha)} \qquad \text{and} \qquad F_B = \frac{\alpha \cdot W}{(2 + \alpha)}$$

- Note that if we set $\alpha=0$, in effect removing the middle support, we obtain what we obtained before - *indeterminate solution (to equilibrium) #1.*

- Note that if we set $\alpha=1.0$, so that all three springs have the same stiffness, we obtain what we obtained before - *full indeterminate solution, Scenario #1.*

- Note that if we let α be a very, very large number, then the forces in the springs at the ends become very, very small relative to the force in the spring at mid-span. In effect we have removed them. (We leave the stability of this situation to a later chapter).

Full Indeterminate solution, Scenario #3

We can play around with the relative values of the stiffness of the three springs all day if we so choose. While not wanting to spend all day in this way, we should at least consider one scenario in which we loose the symmetry, in which case the springs experience different deformations.

Let us take the stiffness of the spring at the left end equal to the stiffness of the spring at midspan, but now set the stiffness of the spring at the right equal to but a fraction of the former;

$$F_A = k \cdot \delta_A$$

That is, we take
$$F_B = k \cdot \delta_B$$

$$F_C = \alpha k \cdot \delta_C$$

Clearly we have lost our symmetry. We need to reconsider compatibility of deformation, considering how the displacements of the three springs must be related.

The figure at the right is *not* a free body diagram. It is a new diagram, simpler in many respects than a free body diagram. It is a picture of the displaced structure, rather a picture of how it might possibly displace.

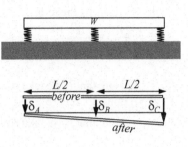

"Possibilities" are limited by our requirement that the block remain all in one piece and *rigid*. This means that the points representing the locations of the ends of the springs, at their junctions with the block, in the displaced state *must all lie on a straight line*.

The figure shows the *before* and *after* loading states of the system.

There is now a rotation of the block as well as a vertical displacement[7]. Now, we know that it takes only two points to define a straight line. So say we pick δ_A and δ_B and pass a line through the two points. Then, if we extend the line to the length of the block, the intersection of a vertical line drawn through the end at C in the undeflected state and this extended line will define the displacement δ_C.

In fact, from the geometry of this displaced state, chanting "...similar triangles...", we can claim

$$\frac{(\delta_B - \delta_A)}{(L/2)} = \frac{(\delta_C - \delta_A)}{L} \qquad \text{or} \qquad \delta_B = \frac{1}{2} \cdot (\delta_C + \delta_A)$$

This second equation shows that the midspan displacement is the mean of the two end displacements.

This is our *compatibility* condition. It holds *irrespective of our choice of spring stiffness*. It is an *independent requirement*, independent of equilibrium as well. It is a consequence of our assumption that the block is rigid.

Now, with our assumed constitutive relations, we find that the forces in the springs may be written in terms of the displacements as follows.

$$F_A = k \cdot \delta_A$$

$$F_B = k \cdot \frac{1}{2} \cdot (\delta_C + \delta_A) \qquad \text{where we have eliminated } \delta_B \text{ from our story.}$$

$$F_C = \alpha k \cdot \delta_C$$

Equilibrium, expressed in terms of the two displacements, δ_A and δ_C. gives:

$$\delta_A + \frac{1}{2} \cdot (\delta_C + \delta_A) + \alpha \delta_C = W/k \qquad \text{and} \qquad \frac{(\delta_C + \delta_A)}{4} + \alpha \delta_C = W/(2k)$$

---

7. We say the system now has *two degrees of freedom*.

The solution to these is:

$$\delta_A = \frac{2W}{k} \cdot \frac{\alpha}{(1+5\alpha)} \qquad \delta_B = \frac{W}{k} \cdot \frac{(1+\alpha)}{(1+5\alpha)} \qquad \text{and} \qquad \delta_C = \frac{2W}{k} \cdot \frac{1}{(1+5\alpha)}$$

- Note that if we take $\alpha = 1.0$, we again recover the symmetric solution $\quad \delta_A = \delta_B = \delta_C \quad$ and $\quad F_A = F_B = F_C = W/3$

- Note that if we take $\alpha = 0$ we obtain the interesting result $\delta_A = 0 \qquad \delta_B = W/k \qquad \delta_C = 2 \cdot W/k$ which means that the block pivots about the left end. And the midspan spring carries all of the weight of the block! $\quad F_A = F_C = 0 \qquad F_B = W$

- And if we let α get very, very large...(see the problem at the end of the chapter.

Full Indeterminate solution, Scenario #4

As a final variation on this problem, we relinquish our claim that the block is rigid. Say it is not made of stone, but of some more flexible, structural material such as aluminum, or steel, or wood, or even glass. We still assume that the weight is uniformly distributed over its length.

We will, however, assume the spring stiffness are of a special form in order to obtain a relatively simple problem formulation and resolution. We take the end springs as infinitely stiff, as rigid. They deflect not at all. In effect we support the block at its ends by pins. The stiffness of the spring at midspan we take as k.

Our picture of the geometry of deformation must be redrawn to allow for the relative displacement of points, any two points, in the block.

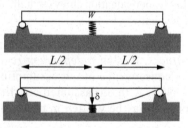

We again, assuming the block is uniform along its length, can claim symmetry. We sketch the deflected shape accordingly.

Equilibrium remains as before. But now we must be concerned with the constitutive relations for the beam!

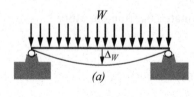

(a)

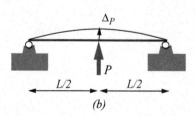

(b)

Forget the spring for a moment. Picture the beam as subjected to two different loadings: The first, a uniformly distributed load, figure (a); the second, a concentrated load P at midspan, as shown in figure (b), due to the presence of the spring.

Let the deflection at midspan due to first loading condition, W, uniformly distributed load, be designated by Δ_W. Take it from me that we can write

$$W = k_W \cdot \Delta_W$$

that is, the deflection grows linearly with the weight[8]. Here k_W is the stiffness, relating the midspan deflection to the total weight of the block.

Let the deflection at midspan due to the second loading condition, the concentrated load, P, be designated by Δ_P. Take it from me that we can write $P = k_P \cdot \Delta_P$ where we will, in time, identify P as the force due to the compression of the spring.

Now, for *compatibility of deformation*, the actual deflection at midspan will be the difference of these two deflections: If we take downward as positive we have

$$\delta = \Delta_W - \Delta_P$$

where δ is the net downward displacement at midspan and hence, the actual compressive displacement of the spring. Putting this in terms of the spring force and the weight W, and the force P, we have:

$$F_B / k = W / k_W - P / k_P \quad \text{But P is just } F_B. \text{ So we have } F_B / k = W / k_W - F_B / k_P$$

We solve this now for the force in the spring in terms of the total weight, W, which we take as given and obtain

$$F_B = \frac{k_P \cdot k}{k_W} \cdot \frac{W}{(k + k_P)}$$

This simplifies if you accept the fact that the ratio of the stiffness, k_P / k_W, is known. Take it from me that this ratio is 5/8.

With this we can write $\qquad F_B = (5/8) \cdot \dfrac{W}{(1 + k_P / k)}$

Then from force equilibrium we obtain: $\qquad F_A = F_C = \dfrac{1}{2} \cdot \dfrac{(3/8 + k_P/k)}{(1 + k_P/k)} \cdot W$

- Note that if we let the stiffness, k, of the spring get very, very large, we get $F_A = F_C = (3/16) \cdot W$ while $F_B = (5/8) \cdot W$ In effect, we have replaced the midspan spring with a rigid, pin support and these are the reaction forces at the supports.

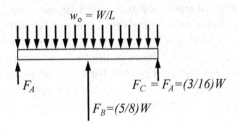

$$w_o = W/L$$

F_A

$F_C = F_A = (3/16)W$

$F_B = (5/8)W$

---

8. We study, and construct expressions for the displacement distribution of beams due to various loading conditions in a later chapter.

- Note how, knowing the reaction forces, we could go on and draw the shear-force and bending moment diagram.

That's enough variations on single problem. We turn now to a second exercise, an indeterminate problem again, to see the power of our three principles of analysis.

Exercise 5.2

A rigid carton, carrying fragile contents (of negligible weight), rests on a block of foam and is restrained by four elastic cords which hold it fast to a truck-bed during transit. Each cord has a spring stiffness k_{cord} = 25 lb/in.; the foam has eight times the spring stiffness, k_{foam} = 400 lb/in.

The gap Δ, in the undeformed state, i.e., when the cords hang free, is 1.0 in.[9] Show that when the carton is held down by the four cords that each of the cords experiences a tensile force of 20 lb.

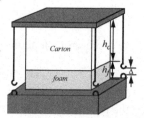
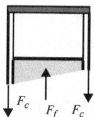

We begin by making a cut through the body to get at the internal forces in the cords and in the foam. We imagine the cords hooked to the floor; the system in the deformed state. We cut through the foam and the cords at some quite arbitrary distance up from the floor. Our isolation is shown at the right.

Force equilibrium gives, noting there are four cords to take into account:

$$F_f - 4F_c = 0$$

Here we model the system as capable of motion in the vertical direction only. The internal reactive force in the foam is taken as uniformly distributed across the cut. F_f is the resultant of this distribution. Consistent with this, the foam, like the cords, is taken as a uniaxial truss member, like a linear spring.

Observe that the foam will compress, the cords will extend. Note that I seemingly violate my convention for assumed direction of positive truss member forces in that I take a *compressive* force in the foam as positive. I could argue that this is not a truss; but, no, the real reason for proceeding in this way is to make full use of our physical insight in illustrating the new requirement of compatibility of deformation. We can be quite confident that the foam will compress and the

---

9. It is not necessary to state that we neglect the weight of the carton if we work from the deformed state with the foam deflected some due to the weight alone. This is ok as long as the relationship between force and deflection is linear, which we assume to be the case.

cords extend. In other instances to come, the sign of the internal forces will **not** be so clear. Careful attention must then be given to the convention we adopt for the positive directions of *displacements*, as well as forces. Note also that there exists *no* externally applied forces yet internal forces exist and must satisfy equilibrium. We call this kind of system of internal forces *self-equilibrating*.

We see we have but **one equation** of equilibrium, **yet two unknowns**, the internal forces F_c and F_f. The problem is *statically, or equilibrium indeterminate*.

We now call upon the new requirement of *compatibility of deformation* to generate another required relationship. In this we designate the compression of the foam δ_f; its units will be *inches*. We designate the extension of the cords δ_c; it too will be measured in *inches*.

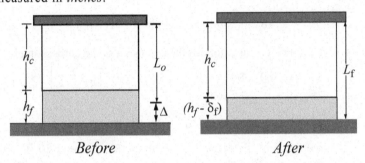

Before   *After*

Compatibility of deformation is a statement relating these two measures. In fact their sum must be, Δ, the original gap. We construct this statement as follows:

The original length of the cord is

$$L_o = h_f + h_c - \Delta \quad \text{while its final length is} \quad L_f = (h_f - \delta_f) + h_c$$

The *extension* of the cord is the difference of these: $\quad \delta_c = L_f - L_O = -\delta_f + \Delta$

Compatibility of Deformation then requires

$$\boxed{\delta_c + \delta_f = \Delta}$$

Only if this is true will our structure remain *all together now* as it was before fastening down.

Here is **a second equation**, but look, we have introduced **two more unknowns**, the compression of the foam and the extension of the cords. It looks like we are making matters worse! Something more must be added, namely we must relate the internal forces that appear in equilibrium to the deformations that appear in compatibility. This is done through two *constitutive equations*, equations whose form and factors depend upon the material out of which the cord and foam are *constituted*. In this example we have modeled both the foam and the cords as *linear springs*. That is, we write

$$F_c = k_c \cdot \delta_c \qquad where \qquad k_c = 25(lb/in)$$
$$and$$
$$F_f = k_f \cdot \delta_f \qquad where \qquad k_f = 400(lb/in)$$

These last are **two more equations**, but **no more unknowns**. Summing up we see we now have four *linearly independent* equations for the four unknowns, — the two internal forces and the two measure of deformation.

There are various ways to solve this set of equations; I first write δ_f in terms of δ_c using compatibility, i.e. $\delta_f = \Delta - \delta_c$ then express both unknown forces in terms of δ_c.

$$F_c = k_c \cdot \delta_c \qquad and \qquad F_f = k \cdot (\Delta - \delta_c)$$

Equilibrium then yields a single equation for the extension of the cords, namely

$$k_f \cdot (\Delta - \delta_c) - 4k\delta_c = 0 \quad so \quad \delta_c = \Delta \cdot \left[\frac{k_f}{4k_c + k_f} \right]$$

and we find the tension in the cords, F_c to be:

$$F_c = k_c \cdot \delta_c = 20 \text{ lb.}$$

The compressive force in the foam is four times this, namely 80 *lb*, since there are four cords. Finally, we find that the extension of the cords and the compression of the foam are

$$\delta_c = 0.8 \text{ in.} \qquad and \qquad \delta_f = 0.2 \text{ in.}$$

which sum to the original gap, Δ.

This simple exercise[10] captures all of the major features of the solution of statically indeterminate problems. We see that we must contend with **three requirements: Static Equilibrium, Compatibility of Deformation, and Constitutive Relations**. A less fancy phrasing for the latter is *Force-Deformation Equations*. We turn now to a third exercise which includes truss members under uniaxial loading.

---

10. *Simplicity* is not meant to imply that the exercise is not without practical importance or that it is a simple matter to conjure up all the required relationships: If I were to throw in a little dash of the dynamics of a single degree of freedom, Physics I, Differential Equations, mass-spring system I could start designing cord-foam support systems for the safe transport of fragile equipment over bumpy roads. More to come on this score.

Exercise 5.3

I know that the tip deflection at the end C of the structure — made of a rigid beam ABC of length L= 4m, and two 1020CR steel support struts, DB and EB, each of cross sectional area A and intersecting at a= L/4 — when supporting an individual weighing 800 Newtons is 0.5mm. What if I suspend more individuals of the same weight from the point C; when will the structure collapse?

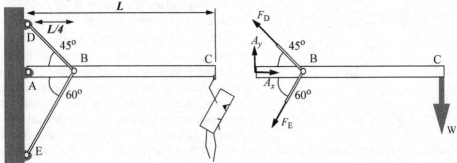

Here is a problem statement which, when you approach the punch line, prompts you to suspect the author intends to ask some ridiculous question, e.g., "What time is it in Chicago?" No matter. We know that if it's in this textbook it is going to require a free-body diagram, application of the requirements of static equilibrium, and now, compatibility of deformation and constitutive equations. So we proceed. I start with equilibrium, isolating the *rigid* bar, *ABC*.

Force Equilibrium:

$$A_x - F_D \cdot \cos 45^\circ - F_E \cdot \cos 60^\circ = 0$$

$$A_y + F_D \cdot \sin 45^\circ - F_E \cdot \sin 60^\circ - W = 0$$

Moment equilibrium (positive ccw), about point[11] *A* yields

$$F_D \cdot \sin 45^\circ \cdot (L/4) - F_E \cdot \sin 60^\circ \cdot (L/4) - WL = 0$$

These are three equations for the four unknowns, A_x, A_y, F_D, and F_E. The structure is redundant. We could remove either the top or the bottom strut and the remaining structure would support an end load – not as great an end load but still some significant value.

---

11. Member *ABC* is *not* a two-force member even though it shows frictionless pins at *A, B* and *C*. In fact it is not a two-force member because it is a three-force member– three forces act at the three pins *(F_D* and *F_E* may be thought of as equivalent to a single resultant acting at *B*). The member must also support an internal bending moment, i.e., over the region *BC* it acts much like a cantilever beam. Note that, while there can be no couple acting at the interface of the frictionless pin and the beam at *B*, there is a bending moment internal to the beam at a section cut through the beam at this point. If you can read this and read it correctly you are mastering the language.

Compatibility of Deformation:

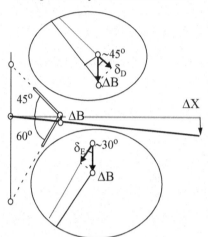

The deformations of member *BD* and member *BE* are related. How they relate is not obvious. We draw a picture, attempting to show the motion of the system from the undeformed state (*W*=0) to the deformed state, then relate the member deformations to the displacement of point *B*.

I have let Δ_B represent the *vertical* displacement of point *B* and Δ_C the vertical displacement at the tip of the *rigid* beam. Because I have said that member *ABC* is *rigid*, there is no horizontal displacement of point *B* or at least none that matters. If the member is elastic, the horizontal displacement should be taken into account in relating the deformations of the two struts. When the member is rigid, there **is** a horizontal displacement of *B* but for *small vertical displacements*, Δ_B, the horizontal displacement is *second order*. For example, if Δ_B/L is of order 10^{-1}, then the horizontal displacement is of order 10^{-2}.

Shown above the full structure is an exploded view of the vertical displacement Δ_B and its relationship to the deformation of member *DB*, the extension δ_D. From this figure I take

$$\delta_D = \Delta_B \cdot \cos 45^o = (\sqrt{2}/2)\Delta_B$$

Shown at the bottom is an exploded view of the vertical displacement and its relationship to the deformation of member *EB*. Taking the measures of deformation as positive in extension, consistent with our convention of taking the member forces as positive in tension, and noting that member *EB* will be in compression, we have

$$\delta_E = -\Delta_B \cdot \cos 30^o = -(\sqrt{3}/2)\Delta_B$$

These two equations relate the deformations of the two struts through the variable Δ_B. We can read them as saying that, for small deflections and rotations, **the extension or contraction of the member is equal to the projection of the displacement vector upon the member.**

Only if these equations are satisfied are these deformations compatible; only then will the two members remain together, joined at point *B*. This is then our requirement of **compatibility of deformation.**

Constitutive Equations:

These are the simplest to write out. We assume the struts are both operating in the elastic region. We have

$$\sigma_{DB} = E \cdot \varepsilon_{DB} \qquad \text{or} \qquad F_D/A = E \cdot \delta_D/(\sqrt{2} \cdot L/4)$$

and

$$\sigma_{EB} = E \cdot \varepsilon_{EB} \qquad \text{or} \qquad F_E/A = E \cdot \delta_{DE}/(L/2)$$

where I have used the geometry to figure the lengths of the two struts.

Now let's go back and see what was given, what was wanted. We clearly are interested in the forces in the two struts, the two *F's*; more precisely, we are interested in the stresses engendered by the end load *W*, for if either of these stresses reaches the *yield strength* for 1020CR steel we leave the elastic region and must consider the possibility of collapse of our structure. These forces, in turn, depend upon the member deformations, the δ's which, in turn depend upon the vertical deflection at *B*, Δ_B.

We can think of the problem, then, as one in which there are five unknowns. We see that we have *seven* equations available, but note we have the horizontal and vertical components of the reaction force at *A* as unknowns too, so everything is in order. In fact we only need work with five of the seven equations because A_x and A_y appear only in the two force equilibrium equations. The wise choice of point *A* as our reference point for moment equilibrium enables us to proceed without worrying about these two relations. [12] I first express the member forces in terms of Δ_B using the constitutive and compatibility relations. I obtain

$$F_D = 2 \cdot (AE/L) \cdot \Delta_B \qquad \text{and} \qquad F_E = -\sqrt{3} \cdot (AE/L) \cdot \Delta_B$$

where the negative sign indicates that member *EB* is in compression. Note that the magnitude of the tensile load in *DB* is greater than the compression in member *EB*. Now substituting these for the forces as they appear in the equation of moment equilibrium I obtain the following relationship between the end load *W* and the vertical displacement of point *B*, namely:

$$W = \left(\frac{3 + 2\sqrt{2}}{8}\right) \cdot (AE/L) \cdot \Delta_B$$

This relationship is worth a few words: It relates the vertical force, *W*, at the point *C*, at the end of the rigid beam, to the vertical displacement, Δ_B, at *another* point in the structure. The factor of proportionality can be read as a stiffness, k, like that of a linear spring. Note that the dimensions of the factor, (AE/L) are just force per length.

---

12. If we were concerned, as perhaps we should be, with the integrity of the fastener at *A* we would solve these two equations to determine the reaction force.

Now I can use this, and the observation that when $W = 800N$, the tip deflection is 0.5mm to obtain an expression for the factor (AE/L). But first I have to relate Δ_B to the tip deflection $\Delta_C = .5$mm. For this I return to my sketch of the deformed geometry and note, if (and only if) *ABC* is *rigid* then the two deflections are related, "through similar triangles" by $\Delta_B = \Delta_C/4$

This then yields

$$(AE/L) = \frac{8W}{\left[\left(\frac{\Delta_C}{4}\right)\cdot(3+2\sqrt{2})\right]} = 8.76 \times 10^6 N/m$$

Now with the length given as 4 *meters* and E the elastic modulus, found from a table in Chapter 7, to be $200\times 10^9 \ N/m^2$, we find that the cross sectional area is

$$A = 1.76 \times 10^{-4} m^2 = 176 \ mm^2$$

If the struts were solid and circular, this implies a 15 *mm* diameter.

The stress will be bigger in member *DB* than member *EB*. In fact, from our expression above for F_D, we have $\sigma_D = 2\cdot E\cdot(\Delta_B/L)$ This, evaluated for the particular deflection recorded with one individual supported at C and taking E as before yields $\sigma_D = 12.6\times 10^6 \ N/m^2$.

If we *idealize* the constitutive behavior as *elastic-perfectly plastic* and take the yield strength as $600\times 10^6 \ N/m^2$, we conclude that we could suspend *forty-seven individuals*, each a hefty weight of 800 *Newtons* before the onset of yield in the strut *BD*, before collapse becomes a possibility.[13] But will it collapse at that point? No, not in this idealized world anyway. Member *EB* has yet to reach its yield strength; once it does, then the structure, again in this idealized world, can support no further increase in end load without infinite deflection and deformations of the struts.

---

13. This is a strange kind of problem – using the observed displacement under a known load to calculate, to *back-out* the cross sectional areas of the struts. The ordinary, *politically correct* textbook problem would specify the area and everything else you needed (but not a wit more) and ask you to determine the tip displacement. But nowadays machines are given that kind of straight-forward problem to solve. A more challenging kind of dialogue in Engineering Mechanics – common to *diagnostic* situations where your structure is not behaving as expected, when something goes wrong, deflects too much, fractures too soon, resonates at too low a frequency, and the like – demands that you construct different scenarios for the observed behavior, e.g., (did it deflect too much because the top strut exceeded the yield strength?), and test their validity. The fundamental principles remain the same, the language is the same language, but the context is much richer; it places greater emphasis upon your ability to formulate the problem, to construct a story that explains the system's behavior. Often, in these situations, you will not have, or be able to obtain, full or complete information about the structure. In this case, *backing out* the area of the struts might be just one step in diagnosing and explaining the observed and often mystifying, behavior.

5.2 Matrix analysis of Truss Structures - Displacement Formulation

The problems and exercises we have assigned to date have all been amenable to solution *by hand*. We now consider a method of analysis especially well suited for truss structures that takes advantage of modern computer power and allows us to address structures with many nodes and members. Our aim is to find all internal member forces and all nodal displacements given some external forces applied at the nodes. In this, we make use of a *displacement formulation* of the problem; the unknowns of the final set of equations we give the computer to solve are the components of the node displacements. We use matrix notation in our formulation as an efficient and concise way to represent the large number of equations that enter into our analysis.

These equations will account for i) equilibrium of internal member forces and external forces applied at the nodes, ii) the force/deformation behavior of each member and iii) compatibility of the extension and contraction of members with the displacements at the nodes. If the structure is statically determinate and we seek only to determine the forces in the truss members, we need only consider the first of these three sets of equations. If, however, we want to go on and determine the displacements of the nodes as well, we then must consider the full set of relations. If the truss is *statically indeterminate*, if it has redundant members, then we must always, by necessity, consider the deformations of members and displacements of nodes as well as satisfy the equilibrium equations.

To illustrate the *displacement method* we do two examples, one that could be done by hand, the second that is more efficiently done by computer.

Exercise 5.3– The members of the redundant structure shown below have the same cross sectional area and are all made of the same material. Show

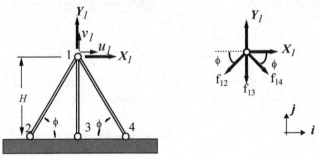

that the equations expressing force equilibrium of node #1 in the x and y directions, when phrased in terms of the unknown displacements of the node, u_1, v_1, take the form

$$2(AE/L) \cdot (\sin\phi \cdot \cos^2\phi) \cdot u_1 = X_1 \qquad and \qquad (AE/L) \cdot (1 + 2\sin^3\phi) \cdot v_1 = Y_1$$

Note: We let X and Y designate the x,y components of the applied force at the node, while u_1 and v_1 designate the corresponding components of displacement of the node.

Equilibrium. with respect to the undeformed configuration, of node #1

$$-f_{12}\cos\phi + f_{14}\cos\phi + X_1 = 0 \qquad and \qquad -f_{12}\sin\phi - f_{13} - f_{14}\sin\phi + Y_1 = 0$$

These are two equations in three unknowns as we expected since the structure is redundant. In matrix notation they take the form:

$$\begin{bmatrix} \cos\phi & 0 & -\cos\phi \\ \sin\phi & 1 & \sin\phi \end{bmatrix} \begin{bmatrix} f_{12} \\ f_{13} \\ f_{14} \end{bmatrix} = \begin{bmatrix} X_1 \\ Y_1 \end{bmatrix}$$

Compatibility of deformation of the three members is best viewed from the following perspective: Imagine an arbitrary *displacement* of node #1, a vector with two scalar x,y components $u = u_1 i + v_1 j$ where, as usual, i, j are two unit vectors directed along the x,y axes respectively.

We take as a measure of the **member deformation**, say of member 1-2, **the projection of the displacement upon the member.** We must be careful to take account if the member extends or contracts. In the second example we show a way to formally do this bit of accounting. Here we rely upon a sketch. Shown below is member 1-2 and an arbitrary node displacement **u** drawn as if both of its components were positive.

$$t_{12} = \cos\phi\, i + \sin\phi\, j$$
$$u = u_1 i + v_1 j$$
$$\delta_{12} = u \bullet t_{12}$$

The projection upon the member is given by the scalar, or *dot* product

$$\delta_{12} = u \bullet t_{12} \quad \text{where} \quad t_{12} = \cos\phi\, i + \sin\phi\, j$$

is a unit vector directed as shown, along the member in the direction of a positive extension. We thus obtain three equations relating the three *member deformations* to the two *nodal displacement*[14]

$$\delta_{12} = u_1 \cos\phi + v_1 \sin\phi$$
$$\delta_{14} = -u_1 \cos\phi + v_1 \sin\phi$$
$$\delta_{13} = v_1$$

If they are satisfied, we can rest assured that our structure remains all of one piece in the deformed configuration. In matrix notation, they take the form:

---

14. Note that the horizontal displacement component, u_1, engenders no elongation or contraction of the middle, vertical member. This is a consequence of our assumption of small displacements and rotations.

$$\begin{bmatrix} \delta_{12} \\ \delta_{13} \\ \delta_{14} \end{bmatrix} = \begin{bmatrix} \cos\phi & \sin\phi \\ 0 & 1 \\ -\cos\phi & \sin\phi \end{bmatrix} \begin{bmatrix} u_1 \\ v_1 \end{bmatrix}$$

Keeping count, we now have five scalar equations for eight unknowns, the three member forces, the three member deformations, and the two nodal displacements. We turn now to the three...

Force-Deformation relations are the usual for a truss member, namely

$$f_{12} = (AE/L_{12}) \cdot \delta_{12} \qquad f_{13} = (AE/L_{13}) \cdot \delta_{13} \qquad f_{14} = (AE/L_{14}) \cdot \delta_{14}$$

The lengths may be expressed in terms of H, e.g.,

$$L_{12} = L_{14} = H/\sin\phi \qquad and \qquad L_{13} = H$$

These, in matrix notation, take the form

$$\begin{bmatrix} f_{12} \\ f_{13} \\ f_{14} \end{bmatrix} = \begin{bmatrix} \left(\dfrac{AE\sin\phi}{H}\right) & 0 & 0 \\ 0 & \left(\dfrac{AE}{H}\right) & 0 \\ 0 & 0 & \left(\dfrac{AE\sin\phi}{H}\right) \end{bmatrix} \begin{bmatrix} \delta_{12} \\ \delta_{13} \\ \delta_{14} \end{bmatrix}$$

Displacement Formulation. first expressing the member forces in term of the nodal displacements using compatibility and the force-deformation equations, (in matrix notation)

$$\begin{bmatrix} f_{12} \\ f_{13} \\ f_{14} \end{bmatrix} = \begin{bmatrix} \left(\dfrac{AE\sin\phi}{H}\right) & 0 & 0 \\ 0 & \left(\dfrac{AE}{H}\right) & 0 \\ 0 & 0 & \left(\dfrac{AE\sin\phi}{H}\right) \end{bmatrix} \begin{bmatrix} \cos\phi & \sin\phi \\ 0 & 1 \\ -\cos\phi & \sin\phi \end{bmatrix} \begin{bmatrix} u_1 \\ v_1 \end{bmatrix}$$

then substitute for the forces in the equilibrium equations. We have, again continuing with our matrix representation:

$$\begin{bmatrix} \cos\phi & 0 & -\cos\phi \\ \sin\phi & 1 & \sin\phi \end{bmatrix} \cdot \begin{bmatrix} \left(\dfrac{AE\sin\phi}{H}\right) & 0 & 0 \\ 0 & \left(\dfrac{AE}{H}\right) & 0 \\ 0 & 0 & \left(\dfrac{AE\sin\phi}{H}\right) \end{bmatrix} \cdot \begin{bmatrix} \cos\phi & \sin\phi \\ 0 & 1 \\ -\cos\phi & \sin\phi \end{bmatrix} \cdot \begin{bmatrix} u_1 \\ v_1 \end{bmatrix} = \begin{bmatrix} X_1 \\ Y_1 \end{bmatrix}$$

Carrying out the matrix products yields a set of two scalar equations for the two nodal displacements:

$$\left(\frac{AE}{H}\right)\begin{bmatrix} 2\sin\phi(\cos\phi)^2 & 0 \\ 0 & 1+2\sin^3\phi \end{bmatrix}\begin{bmatrix} u_1 \\ v_1 \end{bmatrix} = \begin{bmatrix} X_1 \\ Y_1 \end{bmatrix}$$

These are the *equilibrium equations in terms of displacements.* They can be easily solved since they are *uncoupled,* that is each can be solved independently for one or the other of the nodal displacements. The symmetry of the structure is the reason for this happy outcome.   This becomes clear when we write them out according to our more ordinary habit and obtain what we sought to show:

$$(AE/H) \cdot (2\sin\phi\cos\phi^2)u_1 = X_1$$

and

$$(AE/H) \cdot (1+2\sin\phi^3)v_1 = Y_1$$

Unfortunately this decoupling doesn't occur often in practice as a second example shows. We turn to that now, a more complex structure, which in a first instance we take as statically determinate.

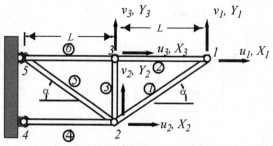

Now consider the truss structure shown above. Although this system is more complex than the previous example in that it has more *degrees of freedom* – six scalar nodal displacements versus two for the simpler truss – the structure is less complex in that it is statically determinate; there are no redundant or unnecessary members; remove any member and the structure would collapse.

We first develop a set of equilibrium equations by isolating each of the free nodes and requiring the sum of all forces, internal and external, to vanish. In this, lower case *f* will represent member forces, assumed to be positive when the member is in tension, and upper case *X* and *Y*, the *x* and *y* components of the externally applied forces.

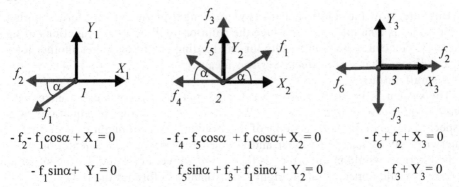

$$-f_2 - f_1\cos\alpha + X_1 = 0 \qquad -f_4 - f_5\cos\alpha + f_1\cos\alpha + X_2 = 0 \qquad -f_6 + f_2 + X_3 = 0$$

$$-f_1\sin\alpha + Y_1 = 0 \qquad f_5\sin\alpha + f_3 + f_1\sin\alpha + Y_2 = 0 \qquad -f_3 + Y_3 = 0$$

These six equations for the six unknown member forces can be put into matrix form

$$\begin{bmatrix} \cos\alpha & 1 & 0 & 0 & 0 & 0 \\ \sin\alpha & 0 & 0 & 0 & 0 & 0 \\ -\cos\alpha & 0 & 0 & 1 & \cos\alpha & 0 \\ -\sin\alpha & 0 & -1 & 0 & -\sin\alpha & 0 \\ 0 & -1 & 0 & 0 & 0 & 1 \\ 0 & 0 & 1 & 0 & 0 & 0 \end{bmatrix} \begin{bmatrix} f_1 \\ f_2 \\ f_3 \\ f_4 \\ f_5 \\ f_6 \end{bmatrix} = \begin{bmatrix} X_1 \\ Y_1 \\ X_2 \\ Y_2 \\ X_3 \\ Y_3 \end{bmatrix}$$

We could, if we wish at this point, solve this system of six linear equations for the six unknown member forces *f*. If you are so inclined you can apply the methods you learned in your mathematics courses about existence of solutions, about solving linear systems of algebraic equations, and verify for yourself that indeed a *unique* solution does exist. And given enough of your spare time, I wager you could actually carry through the algebraic manipulations and obtain the solution. But our purpose is not to burden you with ordinary menial exercise but rather to show you how to formulate the problem for computer solution. We will let it do the menial and mundane work.

Something is lost, something is gained when we turn to the machine to help solve our problems. The expressions you would obtain by hand for the internal forces would be explicit functions of the applied forces and the parameter α. For example, the second equation alone gives

$$f_1 = Y_1 / (\sin\alpha)$$

The computer, on the other hand, would produce, using the kinds of software common in industry, a solution for specific *numerical* values of the member forces if provided with a specific, numerical value for α and specific numerical values for the externally applied forces at the nodes as input. Of course the computer does this very fast, compared to the time it would take you to produce a solution by hand. And, if need be, with the machine you can make many runs and discover how your results vary with α.

But note: How the solution changes with changes in the external forces applied at the nodes is a simpler matter: since the solutions will be *linear* functions of the X and Y's you can scale your results for one loading condition to get another load-ing condition. That's what *linear systems* means.

A small detour:

The system is linear because we assumed that the structure experiences only *small displacements and rotations.* We wrote our equilibrium equations with respect to the *undeformed geometry* of the structure. If we thought of the structure otherwise, say as made of rubber and allowed for large displacements, our free-body diagrams would be incorrect as they stand above. For example, the situation at node 3 would appear as at the right rather than as before (at the left)

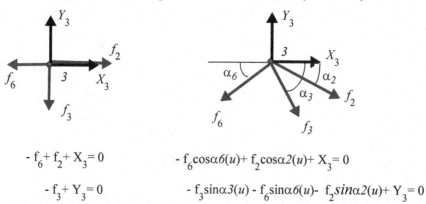

$$-f_6 + f_2 + X_3 = 0$$

$$-f_3 + Y_3 = 0$$

$$-f_6 \cos\alpha 6(u) + f_2 \cos\alpha 2(u) + X_3 = 0$$

$$-f_3 \sin\alpha 3(u) - f_6 \sin\alpha 6(u) - f_2 \sin\alpha 2(u) + Y_3 = 0$$

and our equilibrium equations would now have the more complex form shown.

In these, the alpha's will be unknown functions of all the nodal displacements, for example α_2 will depend upon the displacement of node 3 relative to node 1. We say that the equilibrium equations depend upon the displacements.

But the displacements are functions of the extensions and contractions of the members. These, in turn, are functions of the forces in the members which means that the equations of equilibrium are no longer *linear*. The entries in our square matrix, the coefficients of the unknown forces in our system of six equilibrium equations, depend upon the member forces themselves.

Fortunately, although we did so in our introductory exercise in Chapter 1, you will not be asked to consider large deformations and rotations. The reason is that most structures *do not* experience large deflections and rotations. If they do they are probably in the process of disintegration and failure. Indeed, eventually we will entertain a discussion of *buckling* which ordinarily, though not always, is a mode of failure. We leave, then, the study of such complex, but interesting, modes of behavior to other scholars. Our detour is complete; we return now to more ordinary behavior.

Out in the so-called *real* world, where truss structures span canyons, support aerospace systems, and have hundreds of nodes and members, complexity requires the use of the computer. Imagine a *three-dimensional* truss with 100 nodes. Our linear system of equilibrium equations would number 300; we say that the system has 300 *degrees of freedom*. That is, 300 displacement components are required to fully specify the deformed configuration of the structure. But that is not the end of it: if the structure includes redundant members and hence is statically indeterminate, other equations which relate the member forces to member deformations and still others relating member deformations to node displacements must be written down and solved together with the equilibrium equations. You could still, theoretically, solve all of these hundreds of equations by hand but if you want to remain industrially competitive, if you want to win the bid, you will need the services of a computer.

To illustrate how our system is complicated by adding a redundant member, we connect nodes 3 and 4 with an additional member, number 7 in the figure.

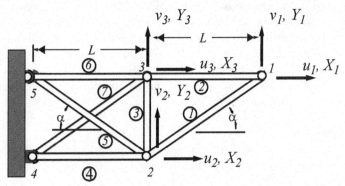

The number of linearly independent equilibrium equations remains the same, namely six, but two of the equations, expressing horizontal and vertical equilibrium of forces at node 3, now include the additional unknown member force f_7. Leaving to you the task of amending the free-body diagram of node #3, we have

$$-f_6 + f_2 - f_7 \cos\alpha + X_3 = 0$$
$$-f_3 - f_7 \sin\alpha + Y_3 = 0$$

With six equations for seven unknowns our problem becomes *statically indeterminate* or *equilibrium indeterminate* as some would prefer. The difficulty is not in finding a solution; indeed, there are an *infinity* of possible solutions. For example we could choose the force in member six to be equal to zero and then solve for all the other member forces. Or we could choose it to equal X_2 and solve, or 10 *lbs*, or 2000 *newtons*, or 2.3 *elephants*, (just be careful with your units), whatever.

Once having arbitrarily specified the force in member six, or the force in any single member for that matter, the six equations will yield values for the forces in all the remaining six members. The difficulty is not in finding a solution, it is in finding a *unique* solution. The problem is indeterminate.

This unique solution, whatever it is, is going to depend upon the kind of member we add to the structure as member number seven. It will depend upon the material properties and cross-sectional area of this new member; for that matter, it will depend upon the force/deformation behavior of *all* members. If the first six members are made of steel and have a cross-sectional area of ten square inches and member seven is a rubber band, we would not expect much difference in our solution for the forces in the steel members when compared to our original solution for those member forces without member seven. If, on the other hand, the added member is also made of steel and has a comparable cross-sectional area, all bets are off, or rather on. The effect of the new member will be significant; the member forces will be substantially different when compared to the statically determinate solution.

Our strategy for solving the statically indeterminate problem is the same one we followed in the previous exercise: We will express all seven unknown internal forces f in terms of the seven, unknown, member deformations which we will designate by δ. We will then develop a method for expressing the member deformations, the δ's, in terms of the x and y components of nodal displacements u and v. There are six of these latter unknowns. After substitution, we will then obtain our six equilibrium equations in terms of the *six* unknown displacement components. *Voila*, a *displacement formulation*.

Equilibrium

The full set of six equilibrium equations in terms of the seven unknown member forces may be written in matrix form as

$$
\begin{bmatrix}
\cos\alpha & 1 & 0 & 0 & 0 & 0 & 0 \\
\sin\alpha & 0 & 0 & 0 & 0 & 0 & 0 \\
-\cos\alpha & 0 & 0 & 1 & \cos\alpha & 0 & 0 \\
-\sin\alpha & 0 & -1 & 0 & -\sin\alpha & 0 & 0 \\
0 & -1 & 0 & 0 & 0 & 1 & \cos\alpha \\
0 & 0 & 1 & 0 & 0 & 0 & \sin\alpha
\end{bmatrix}
\begin{bmatrix}
f_1 \\ f_2 \\ f_3 \\ f_4 \\ f_5 \\ f_6 \\ f_7
\end{bmatrix}
=
\begin{bmatrix}
X_1 \\ Y_1 \\ X_2 \\ Y_2 \\ X_3 \\ Y_3
\end{bmatrix}
$$

or in a condensed form as: $[A]\{f\} = \{X\}$

Note that the array $[A]$ has six rows and seven columns; there are but six equations for the seven unknown internal forces.

Force-Deformation

We assume that the truss members behave like linear springs and, as before, take the member force generated in deformation of the structure as proportional to their change in length δ. We introduce the symbol k for the expression (AE/L) where A is the member cross-sectional area, L its length, and E its modulus of elasticity. For example, for member number 1, we take

$$f_1 = k_1 \cdot \delta_1 \quad \text{where} \quad k_1 = A_1 E_1 / L_1$$

In matrix form,

$$
\begin{bmatrix} f_1 \\ f_2 \\ f_3 \\ f_4 \\ f_5 \\ f_6 \\ f_7 \end{bmatrix}
=
\begin{bmatrix}
k_1 & 0 & 0 & 0 & 0 & 0 & 0 \\
0 & k_2 & 0 & 0 & 0 & 0 & 0 \\
0 & 0 & k_3 & 0 & 0 & 0 & 0 \\
0 & 0 & 0 & k_4 & 0 & 0 & 0 \\
0 & 0 & 0 & 0 & k_5 & 0 & 0 \\
0 & 0 & 0 & 0 & 0 & k_6 & 0 \\
0 & 0 & 0 & 0 & 0 & 0 & k_7
\end{bmatrix}
\begin{bmatrix} \delta_1 \\ \delta_2 \\ \delta_3 \\ \delta_4 \\ \delta_5 \\ \delta_6 \\ \delta_7 \end{bmatrix}
$$

or again, in condensed form:

$$\boxed{f} = \boxed{k} \cdot \boxed{\delta}$$

Compatibility of Deformation

Taking stock at this point we see we have *thirteen* equations but *fourteen* unknowns; the latter include seven member forces f and seven member deformations δ. In this our final step, we introduce another six unknowns, namely the x and y components of the displacements at the nodes and require that the member deformations be consistent with these displacements. *Seven* equations, one for each member, are required to ensure compatibility of deformation. This will bring our totals to *twenty* equations for *twenty* unknowns and allow us to claim victory.

To relate the δ's to the node displacements we consider an arbitrarily oriented member in its undeformed position, then in its deformed state, a state defined by the displacements of its two end nodes. In the following derivation, bold face type will indicate a vector quantity.

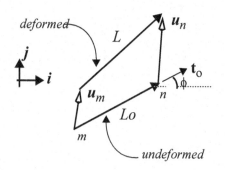

Consider a member with end nodes numbered m and n. Let $\boldsymbol{u}_m$ be the vector displacement of node m. In terms of its x and y scalar components we have:

$$\boldsymbol{u}_m = u_m\boldsymbol{i} + v_m\boldsymbol{j}$$

Member 3

where $\boldsymbol{i}$ and $\boldsymbol{j}$ are unit vectors in the x,y directions. A similar expression may be written for $\boldsymbol{u}_n$.

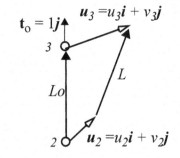

Let $\boldsymbol{L}_0$ be a vector which lies along the member, going from m to n, in its original, undeformed state and $\boldsymbol{L}$ a vector along the member in its displaced, deformed state. Vector addition allows us to write: $\boldsymbol{L}_o + \boldsymbol{u}_n = \boldsymbol{u}_m + \boldsymbol{L}$

Now consider the projection of all of these vector quantities upon a line lying along the member in its original, undeformed state, that is along $\boldsymbol{L}_0$. Let $\boldsymbol{t}_0$ be a unit vector in that direction, directed from m to n.

$$\boldsymbol{t}_o = cos\phi\boldsymbol{i} + sin\phi\boldsymbol{j}$$

The projection of $\boldsymbol{L}_0$ upon itself is just the original length of the member, the magnitude of $\boldsymbol{L}_0$, L_0. The projections of the node displacements are given by the scalar products $\boldsymbol{t}_0 \cdot \boldsymbol{u}_m$ and $\boldsymbol{t}_0 \cdot \boldsymbol{u}_n$. Similarly the projection of $\boldsymbol{L}$ is $\boldsymbol{t}_0 \cdot \boldsymbol{L}$ which we take as approximately equal to the magnitude of $\boldsymbol{L}$. This is a crucial step. It is only legitimate if the member experiences *small rotations*. But note, this is precisely the assumption we made in writing out our equilibrium equations.

Our vector relationship then yields, after projection upon the direction $\boldsymbol{t}_0$ of all of its constituents

$$L - L_o \approx \boldsymbol{u}_n \cdot \boldsymbol{t}_o - \boldsymbol{u}_m \cdot \boldsymbol{t}_o$$

or since the difference of the two lengths is the member's extension, we have

$$\delta = \boldsymbol{u}_n \cdot \boldsymbol{t}_o - \boldsymbol{u}_m \cdot \boldsymbol{t}_o$$

For member 1, for example, carrying out the scalar products we have

$$\delta_3 = v_3 - v_2$$

Note how the horizontal displacement components, the u components, do not enter into this expression for the extension (or compression if $v_2 > v_3$) of member 3. That is, any displacement perpendicular to the member does not contribute to its change its length! This is clearly only approximately true, only true for *small displacements and rotations.*

Similar equations can be written for each member in turn. In some cases, ϕ is zero, in other cases a right angle. The full set of seven compatibility relationships, one for each member, can be written in matrix form as:

$$
\begin{bmatrix} \delta_1 \\ \delta_2 \\ \delta_3 \\ \delta_4 \\ \delta_5 \\ \delta_6 \\ \delta_7 \end{bmatrix}
=
\begin{bmatrix}
\cos\alpha & \sin\alpha & -\cos\alpha & -\sin\alpha & 0 & 0 \\
1 & 0 & 0 & 0 & -1 & 0 \\
0 & 0 & 0 & -1 & 0 & 1 \\
0 & 0 & 1 & 0 & 0 & 0 \\
0 & 0 & \cos\alpha & -\sin\alpha & 0 & 0 \\
0 & 0 & 0 & 0 & 1 & 0 \\
0 & 0 & 0 & 0 & \cos\alpha & \sin\alpha
\end{bmatrix}
\begin{bmatrix} u_1 \\ v_1 \\ u_2 \\ v_2 \\ u_3 \\ v_3 \end{bmatrix}
$$

In condensed form we write $\quad [\delta] = [A]^T \cdot [u]$

where $[A]^T$ is the *transpose* of the matrix appearing in the equilibrium equations $[A]$. The consequence of this seemingly happenstance event will be come clear in the final result.

Equilibrium in terms of Displacement

We now do some substitution to obtain the equilibrium equations in terms of the displacement components at the nodes, all the u's and v's. We first substitute for the member forces f, their representation in terms of the member deformations δ and obtain:

$$ [A] \cdot [k] \cdot [\delta] = [X] $$

Now substituting for the δ column matrix its representation in terms of the node displacements we obtain:

$$ [A] \cdot [k] \cdot [A]^T \cdot [u] = [X] $$

which are the *six* equilibrium equations with the *six* displacement components as unknowns.

The matrix product $[A][k][A]^T$ can be carried out in more spare time. We designate the result by $[K]$ and call it the *system stiffness matrix.* It and all of its elements are shown below: In this, c is shorthand for $\cos\alpha$ and s shorthand for $\sin\alpha$.

$$ [K] = [A] \cdot [k] \cdot [A]^T $$

$$[K] = \begin{bmatrix}
(k_1 c^2 + k_1) & k_1 cs & -k_1 c^2 & -k_1 cs & k_2 & 0 \\
k_1 cs & (k_1 s^2) & -k_1 cs & -k_1 s^2 & 0 & 0 \\
-k_1 c^2 & -k_1 cs & \left(k_4 + k_1 c^2 + k_5 c^2\right) & (k_1 cs - k_5 cs) & 0 & 0 \\
-k_1 cs & -k_1 s^2 & (k_1 cs - k_5 cs) & (k_3 + k_1 s^2 + k_5 s^2) & 0 & -k_3 \\
k_2 & 0 & 0 & 0 & (k_2 + k_6 + k_7 c^2) & k_7 cs \\
0 & 0 & 0 & -k_3 & k_7 cs & (k_3 + k_7 s^2)
\end{bmatrix}$$

[K] is *symmetric* (and will always be!) and, for our example is *six by six*.

The equilibrium equations in terms of displacement are, in condensed form

$$\boxed{[K] \cdot [u] = [X]}$$

This is the set of equations the computer solves given adequate numerical values for

- the material properties including the *Young's modulus* or modulus of elasticity, E, and the member's cross-sectional area A,

- member nodes and their coordinates, from which member lengths may be figured, and subsequently together with the material properties, the member stuffiness, (AE/L), computed,

- the externally applied forces at the nodes,

- specification of any fixed degrees of freedom, i.e., which nodes are pinned.

The computer, in effect, *inverts* the system, or global, stiffness matrix [K], and computes the node displacements u and v given values for the applied forces X and Y. Once the displacements have been found, the deformations can be computed from the compatibility relations. Making use of the force/deformation relations in turn, the deformations yield values for the member forces. All then has been resolved, the solution is complete.

Before ending this section, one final observation. A useful physical interpretation of the elements of the system stiffness matrix is available: In fact, the elements of any column of the [K] matrix can be read as the external forces that are required to produce or sustain a special state of deformation, or system of node displacements – namely a unit displacement corresponding to the chosen column and zero displacements in all other degrees of freedom. This interpretation follows from the rules of matrix multiplication.

A Note on Scaling

It is useful to consider how the solution for one particular structure of a specified geometry and subject to a specific loading can be applied to another structure
of similar geometry and similar loading. By "similar loading" we mean a load
vector which is a scalar multiple of the other. By "similar geometry" we mean a
structure whose member lengths are a scalar multiple of the corresponding member lengths of the other - in which case all angles are preserved.

For similar loading, relative to some reference solution designated by a superscript "*", i.e.,

$$[K] \cdot [u^*] = [X^*]$$

we have, if [X] = β [X*] simply that the displacement vector scales accordingly, that is,
from

$$[K] \cdot [u] = [X] = \beta \cdot [X^*]$$

we obtain [u] = β [u*] .

This is a consequence of the *linear* nature of our system (which, in turn, is a
consequence of our assumption of relatively small displacements and rotations).
What it says is that if you have solved the problem for one particular loading, then
the solution for an infinity of problems is obtained by scaling your result for the
displacements (and for the member forces as well) by the factor β which can take
on an infinity of values.

For similar geometries, we need to do a bit more work. We note first that for
both statically determinate and indeterminate systems, the only way length enters
into our analysis is through the member stiffness, k, where $k_j = AE/L_j$. (We
assume for the moment that the cross-sectional areas and the elastic modulae are
the same for each member). The entries in the matrix [A], and so $[A^T]$ are only
functions of the angles the members make, one with another.

Let us designate some reference geometry, drawn in accord with some reference length scale, by a superscript "*", a reference structure in which the member
lengths are defined for all members, j= 1,n by

$$L^*_j = \beta_j \cdot L^*$$

The force-deformation relations $[f] = [k] \cdot [\delta]$ can then be written

$$[f] = (1/L^*) \cdot [k_\beta] \cdot [\delta]$$

where the elements of the matrix $[k_\beta]$ are given by AE/β_j .

Re-doing our derivation of the equilibrium equations expressed in terms of displacements yields

$$\boxed{\left[K_\beta \right] \cdot \left[u^* \right] = L^* \cdot \left[X \right]}$$

where the stiffness matrix $\left[K_\beta \right]$ is given by

$$\boxed{\left[K_\beta \right] = \left[A \right] \cdot \left[k_\beta \right] \cdot \left[A \right]^T}$$

Note that this is only dependent upon the relative lengths of the members, upon the β_j. Then if we change length scales, say our reference length becomes L, we have to solve

$$\boxed{\left[K_\beta \right] \cdot \left[u \right] = L \cdot \left[X \right]}$$

But the solution to this is the same, in form, as the solution to the "*" problem, differing only by the scale factor L/L^*. Hence, solving the reference problem gives us the solution for an infinite number of geometrically similar structures bearing the same loading. (Note that if the loading is scaled down by the same factor by which the geometry is scaled up, the solution does not change!)

5.3   Energy Methods [15]

We have now all the machinery, concepts and principles we need to solve any truss problem. The structure can be equilibrium indeterminate or determinate. It matters little. The computer enables the treatment of structures with many degrees of freedom, determinate and indeterminate.

But before the computer existed, mechanicians solved truss structure problems. One of the ways they did so was via methods rooted in an alternative perspective - one which builds on the notions of work and energy. We develop some of these methods in this section but will do so based on the concepts and principles we are already familiar with, without reference to energy.

The first method may be used to determine the *displacements of a statically determinate truss structure*. Generalization to indeterminate structures will follow.

---

15. The perspective adopted here retains some resemblance to that found in Strang, G., *Introduction to Applied Mathematics*, Wellesley-Cambridge Press, Wellesley, MA., 1986. I use "Energy Methods" only as a label to indicate what this section is meant to replace in other textbooks on Statics and Strength of Materials.

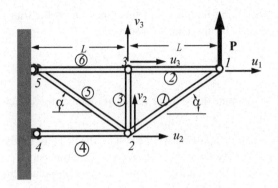

Before proceeding, we review how we might determine the displacements following the path taken in developing the stiffness matrix. We take as an example the statically determinate example of the last section. We simplify the system, applying but one load, **P** in the vertical direction at node 1.

The system is determinate so we solve for the six member forces using the six equations of equilibrium obtained by isolating the structure's three free nodes.

$$- f_2 - f_1 \cos\alpha = 0 \qquad\qquad - f_4 - f_5 \cos\alpha + f_1 \cos\alpha = 0 \qquad\qquad - f_6 + f_2 = 0$$

$$- f_1 \sin\alpha + P = 0 \qquad\qquad f_5 \sin\alpha + f_3 + f_1 \sin\alpha = 0 \qquad\qquad - f_3 = 0$$

These give:

$$f_1 = P/\sin\alpha; \quad f_2 = - P\cos\alpha/\sin\alpha; \quad f_3 = 0; \quad f_4 = 2P\cos\alpha/\sin\alpha;$$
$$f_5 = - P/\sin\alpha; \quad \text{and } f_6 = - P\cos\alpha/\sin\alpha$$

where a positive quantity means the member is in tension, a negative sign indicates compression.

We proceed to determine member deformations, $[\delta]$, from the force/deformation relationships

$$[\delta] = [\, k_{diag} \,]^{-1} [f]$$

that is, from $\delta_1 = f_1/k_1$, $\delta_2 = f_2/k_2$, ... etc; where the k's are the individual member stiffness, e.g.,

$$k_1 = A_1 E_1 / L_1 \quad \text{... etc.}$$

Then, from the compatibility equation relating the six member deformations to the six displacement components at the nodes,

$$[\delta] = [\, A \,]^T [u]$$

we solve this system of six equations for the six displacement components u_1, v_1, u_2, v_2, u_3, v_3. That's it.

A Virtual Force Method

Now consider the alternative method:

We start with the compatibility condition: $\qquad [\delta] = [A]^T [u]$

and take a totally unmotivated step, multiplying both sides of this equation by the transpose of a column vector whose elements may be anything whatsoever;

$$[f^*]^T[\delta] = [f^*]^T[A]^T[u]$$

This arbitrary vector bears an asterisk to distinguish from the vector of member forces acting in the structure.

At this point, the elements of $[f^*]$ could be any numbers we wish, e.g., the price of coffee in the six largest cities of the US (it has to have six elements because the expressions on both sides of the compatibility equation are 6 by 1 matrices). But now we manipulate this relationship, taking the transpose of both sides and write

$$[\delta]^T[f^*] = [u]^T[A][f^*]$$

then consider the vector $[f^*]$ to be *a vector of member forces,* **any** set of member forces that satisfies the equilibrium requirements for the structure, i.e.,

$$[A][f^*] = [X^*]$$

So $[X^*]$ is arbitrary, because $[f^*]$ is quite arbitrary - we can envision many different vectors of applied loads.

With this, our compatibility pre-multiplied by our arbitrary vector, now read as member forces, becomes

$$[\delta]^T[f^*] = [u]^T[X^*] \qquad \text{or} \qquad [u]^T[X^*] = [\delta]^T[f^*]$$

(Note: The dimensions of the quantity on the left hand side of this last equation are displacement times force, or work. The dimensions of the product on the right hand side must be the same).

Now we choose $[X^*]$ in a special way; we take it to be a unit load, a *virtual force*, along a single degree of freedom, all other loads zero. For example, we take

$$[X^*]^T = [\, 0 \ \ 0 \ \ 0 \ \ 0 \ \ 0 \ \ 1 \,]$$

a unit load in the vertical direction at node 3 in the direction of v_3.

Carrying out the product $[u]^T[X^*]$ in the equation above, we obtain just the displacement component associated with the same degree of freedom, v_3 i.e.,

$$v_3 = [\delta]^T[f^*]$$

We can put this last equation in terms of member forces (and member stiffness) alone using the force/deformation relationship and write:

$$v_3 = [f]^T[k]^{-1}[f^*]$$

And that is our special method for determining displacements of a statically determinate truss. It requires, first, solving equilibrium for the "actual" member forces given the "actual" applied loads. We then solve *another* force equilibrium problem - one in which we apply a unit load at the node we seek to determine a displacement component and in the direction of that displacement component.

With the "starred" member forces determined from equilibrium, we carry out the matrix multiplication of the last equation and there we have it.

We emphasize the difference between the two member force vectors appearing in this equation; [f] in plain font, is the vector of the actual forces in structure given the actual applied loads. [f*] with the asterisk, on the other hand, is some, originally arbitrary, force vector which satisfies equilibrium — an equilibrium solution for member forces corresponding to a **unit loading** in the vertical direction at node 3.

Continuing with our specific example, the virtual member forces corresponding to the unit load at node 3 in the vertical direction are, from equilibrium:

$f*_1 = 0$

$f*_2 = 0$

$f*_3 = 1$

$f*_4 = \cos\alpha/\sin\alpha$

$f*_5 = -1/\sin\alpha$

$f*_6 = 0$

With these, and our previous solution for the *actual* member forces, we find

$$v_3 = (P/k_4)(2 \cos\alpha/\sin\alpha)(\cos\alpha/\sin\alpha)+ (P/k_5)/\sin^2\alpha$$

If the members all have the same cross sectional area and are made of the same material, then the ratio of the member stiffness goes inversely as the lengths so $k_5 = \cos\alpha \ k_4$

and, while some further simplification is possible, we stop here.

Virtual Force Method for Redundant Trusses - Maxwell/Mohr Method.

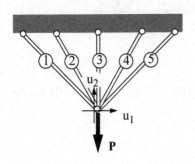

Let's say we have a redundant structure as shown at the left. Now assume we have found all the actual forces, $f_1, f_2,....f_5$, in the members by an alternative method yet to be disclosed (it immediately follows this preliminary remark). The actual loading consists of force components X_1 and X_2 applied at the one free node in directions indicated by u_1 and u_2.

Now say we want to determine the horizontal component of displacement, u_1; Proceeding in accord with our Force Method #1, we must find an equilibrium set of member forces given a unit load applied at the free node in the horizontal direction.

Since the system is redundant, our equilibrium equations number 2 but we have 5 unknowns. The system is indeterminate: it does not admit of a unique solution. It's not that we can't find a solution; the problem is we can find *too many* solutions. Now since our "starred" set of member forces need only satisfy equilibrium, we can arbitrarily set the redundant member forces to zero, or, in effect, *remove them from the structure.* The figure at the right shows one possible choice

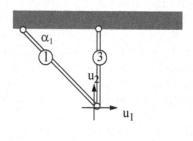

For a unit force in the horizontal direction, we have

$$f_1^* = 1/\cos\alpha_1 \quad \text{and} \quad f_3^* = -1\sin\alpha_1/\cos\alpha_1$$

so the displacement in the horizontal direction, assuming again we have determined the actual member forces, is

$$u_1 = (f_1/k_1)(1/\cos\alpha_1) - (f_3/k_3)(1\sin\alpha_1/\cos\alpha_1)$$

(Note: If the structure is symmetric in member stiffness, k, then this component of displacement, for a vertical load alone, should vanish. This then gives a relationship between the two member forces).

We now develop an alternative method to determine the actual member forces in statically indeterminate truss structures. Consider, for example, the redundant structure shown at the right. We take members 11 and 12 as redundant and write equilibrium in a way that explicitly distinguishes the forces in these two redundant mem-

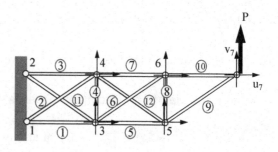

bers from the forces in all the other members. The reasons for this will become clear as we move along.

$$\left[\begin{array}{c|c} A_d & A_r \end{array}\right] \cdot \left[\frac{f_d}{f_r}\right] = \left[X\right]$$

In this, because there are 5 unrestrained nodes, each with two degrees of freedom, the column matrix of external forces, [X], is 10 by 1. Because there are two redundant members, the column matrix [f_r] is 2 by 1. The column matrix of what we take to be "determinate member forces" [f_d] is 10 by 1, i.e., there are a total of

12 member forces. Here, then, are 10 equations for 12 unknowns - an indeterminate system.

The matrix [A$_d$] has 10 rows and 10 columns and contains the coefficients of the 10 [f$_d$]. The matrix [A$_r$], containing coefficients of the 2 [f$_r$], has 10 rows and 2 columns.

Equilibrium can then be re-written

$$[A_d][f_d] + [A_r][f_r] = [X] \quad \text{or} \quad [A_d][f_d] = - [A_r][f_r] + [X]$$

but leave this aside, for now, and turn to compatibility. What we are after is a way to determine the forces in the redundant members without having to explicitly consider compatibility of deformation. Yet of course compatibility must be satisfied, so we turn there now.

The relationship between member deformations and nodal displacements can also be written to explicitly distinguish between the deformations of the "determinant" members and those of the redundant members, that is, the matrix equation [δ] = [A]^T[u] can be written:

$$\begin{bmatrix} \delta_d \\ \hline \delta_r \end{bmatrix} = \begin{bmatrix} A_d^T \\ \hline A_r^T \end{bmatrix} \cdot [u] \quad \text{or} \quad \begin{matrix} [\delta_d] = [A_d]^T \cdot [u] \\ \text{and} \\ [\delta_r] = [A_r]^T \cdot [u] \end{matrix}$$

The top equation on the right is the one we will work with. As in force method #1, we premultiply by the transpose of a column vector (10 by 1) whose elements can be any numbers we wish. In fact, we multiply by the transpose of a general matrix of dimensions 10 rows and *2 columns* - the 2 corresponding to the number of redundant member forces. The reasons for this will become clear soon enough. We again indicate the arbitrariness of the elements of this matrix with an asterisk. We write

$$[f_d*]^T [\delta_d] = [f_d*]^T [A_d]^T[u]$$

In this $[f_d*]^T$ is 2 rows by 10 columns and [δ_d] is 10 by 1.

Now take the transpose and obtain

$$[\delta_d]^T[f_d*] = [u]^T [A_d][f_d*]$$

At this point we choose the matrix [f$_d$*] to be very special; each of the two columns of this matrix (of 10 rows) we take to be a solution to equilibrium. The first column is the solution when

- the external forces [X] are all zero and
- the redundant force in member 11 is taken as a *virtual force* of unity.

The second column is the solution for the determinate member forces when

- the external forces [X] are all zero and
- the redundant force in member 12 is taken as a *virtual force* of unity.

That is, from equilibrium,

$$[A_d] [f_d*] = - [A_r][\ I \ ] \quad \text{where } [\ I \ ] \text{ is the identity matrix.}$$

With this, our compatibility condition becomes

$$[\delta_d]^T[f_d*] = - [u]^T [A_r] \qquad \text{and taking the transpose of this, noting that } [\delta_r] = [A_r]^T[u]$$

we have

$$[f_d*]^T [\delta_d] = - [\delta_r]$$

which gives us the redundant member deformations, $[\delta_r]$, in terms of the "determinate" member deformations, $[\delta_d]$.

But we want the member forces too so we now introduce the member force deformations relations which are simple enough, that is

$$[\delta_r] = [k_r]^{-1} [f_r] \qquad \text{and} \qquad [\delta_d] = [k_d]^{-1} [f_d] \qquad \text{which enables us to write}$$

$$[f_r] = - [k_r][f_d*]^T [k_d]^{-1} [f_d]$$

which, if given the determinate member forces, allows us to compute the redundant member forces.

Substituting, then, back into the equilibrium equations, we can eliminate the redundant forces, expressing the redundant forces in terms of the 10 other member forces, and obtain a system of 10 equations for the 10 unknowns $[f_d]$, namely

$$[\ [A_d] - [A_r][k_r][f_d*]^T [k_d]^{-1} \ ][f_d] = [X]$$

There we have it; a way to determine the member forces in a equilibrium indeterminate truss structure and we don't have to explicitly consider compatibility. What we must do is solve equilibrium several times over; two times to obtain the elements of the matrix $[f_d*]$ in accord with the bulleted conditions stated previously, then, finally, the last equation above, given the applied forces $[X]$.

To go on to determine displacements, we can apply force method #1 - apply a unit load according to the displacement component we wish to determine; use the above two equations to determine all **member forces** (with an asterisk to distinguish them from the actual member forces); then, with the artificial, equilibrium satisfying, "starred" member forces, carry out the required matrix multiplications.

We might wonder how we can get away without explicitly considering compatibility on our way to determining the member forces in an indeterminate truss structure. That we did include compatibility is clear - that's where we started. How does it disappear, then, from view?

The answer is found in one special, mysterious feature of our truss analysis. We have observed, but not proven, that the matrix relating displacements to deformations is the transpose of the matrix relating the applied forces to member forces.

That is, equilibrium gives $\qquad [A][f] = [X]$

While, compatibility gives $\qquad [\delta] = [A]^T[u]$

Now that is bizarre! A totally unexpected result since equilibrium and compatibility are quite independent considerations. (It's the force/deformation relations that tie the quantities of these two domains together). It is this feature which enables us to avoid explicitly considering compatibility in solving an indeterminate problem. Where does it come from? How can we be sure these methods will work for other structural systems?

Symmetry of the Stiffness Matrix - Maxwell Reciprocity

The answer lies in that other domain; that of work and energy. In fact, one can prove that if the work done is to be path independent (which defines an elastic system) then this happy circumstance will prevail.

Consider some quite general truss structure, loaded in the following *two* ways: Let the original, unloaded, state of the system be designated by the subscript "o".

A first method of loading will take the structure to a state "a", where the applied nodal forces $[X_a]$ engender a set of nodal displacements $[u_a]$, then on to state "c" where an *additional* applied set of forces $[X_b]$ engender a set of *additional* nodal displacements, $[u_b]$. Symbolically: $\qquad o \to a \to c = a + b \qquad$ and the work done in following this path may be expressed as[16]

$$Work_{o \to c} = \int_o^c [X]^T \cdot [du] = \int_o^a [X]^T \cdot [du] + \int_a^c [X]^T \cdot [du]$$

and, in that the second integral can be expressed as

$$\int_a^c [X]^T \cdot [du] = \int_a^c [X_a + (X - X_a)]^T \cdot [du] = [X_a]^T \cdot \int_0^b [du] + \int_0^b [X]^T \cdot [du]$$

we have, for this path from o to c:

$$Work_{o \to c} = \int_0^a [X]^T \cdot [du] + \int_0^b [X]^T \cdot [du] + [X_a]^T \cdot [u_b]$$

A second method of loading will take the structure first to state "b", where the applied nodal forces $[X_b]$ engender a set of nodal displacements $[u_b]$, then on to state "c" where an *additional* applied set of forces $[X_a]$ engender a set of *addi-*

---

16. We assume linear behavior as embodied in the stiffness matrix relationship $[X] = [K][u]$.

tional nodal displacements, $[u_a]$. Symbolically: $o \to b \to c = a + b$ And following the same method, we obtain for the work done:

$$Work_{o \to c} = \int_{o}^{a} [X]^{T} \cdot [du] + \int_{o}^{b} [X]^{T} \cdot [du] + [X_b]^{T} \cdot [u_a]$$

Comparing the two boxed equations, we see that for the work done to be path independent we must have

$$[X_a]^{T} \cdot [u_b] = [X_b]^{T} \cdot [u_a]$$

or, with $[X] = [K][u]$

$$[u_a]^{T} \cdot [K] \cdot [u_b] = [u_b]^{T} \cdot [K] \cdot [u_a]$$

from which we conclude that $[K]$, the stiffness matrix, must be symmetric.

Now, since, as derived in a previous section, $[K] = [A] \cdot [k_{diag}] \cdot [A]^{T}$, we see how this must be if work done is to be path independent.

A Virtual Displacement Method.

Given the successful use of equilibrium conditions alone for, not just member forces, but nodal displacements and for indeterminate as well as determinate truss structures, we might ask if we can do something similar using compatibility conditions alone. Here life gets a bit more unrealistic in the sense that the initial problem we pose, drawing on force method #1 as a guide, is not frequently encountered in practice. But it is a conceivable problem - a problem of prescribed displacements. It might help to think of yourself being set down in a foreign culture, a different world, where mechanicians have only reluctantly accepted the reality of forces but are well schooled in displacements, velocities and the science of anything that moves, however minutely.

That is, we consider a truss structure, all of whose displacement components are prescribed, and we are asked to determine the external forces required to give this system of displacements. In the figure at the right, the vectors shown are meant to be the known prescribed displacements. (Node #1 has zero displacement). The task is to find the external forces, e.g., X_3, Y_3, which will produce this deformed

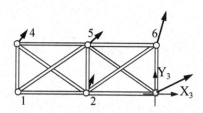

state and be in equilibrium - and we want to do this without considering equilibrium explicitly!

We start with equilibrium[17]:

[X] = [A] [f]

and take a totally unmotivated step, multiplying both sides of this equation by the transpose of a column vector whose elements may be anything whatsoever;

$$[u*]^T[X] = [u*]^T[A] [f]$$

This arbitrary vector bears an asterisk to distinguish it from the vector of actual displacement prescribed at the nodes.

At this point, the elements of [u] could be any numbers we wish, e.g., the price of coffee in the 12 largest cities of the US (it has to have twelve elements because the expressions on both sides of the equilibrium equation are 12 by 1 matrices). But now we manipulate this relationship, taking the transpose of both sides and write

$$[X]^T[u*] = [f]^T[A]^T[u*]$$

then consider the vector [u*] to be *a vector of nodal displacements,* **any** set of nodal displacements that satisfies the compatibility requirements for the structure, i.e.,

$$[A]^T[u*] = [\delta*]$$

So [δ*] is still arbitrary, because [u*] is quite arbitrary - we can envision many different sets of member deformations.

With this, our equilibrium equations, pre-multiplied by our arbitrary vector becomes

$$[X]^T[u*] = [f]^T[\delta*] \quad \text{or} \quad [u*]^T [X] = [\delta*]^T[f]$$

(Note: The dimensions of the quantity on the left hand side of this last equation are displacement times force, or work. The dimensions of the product on the right hand side must be the same).

Now we choose [u*] in a special way; we take it to represent a unit, *virtual* displacement associated with a single degree of freedom, all other displacements zero. For example, we take

$$[u*]^T = [\, 0 \ \ 0 \ \ 0 \ \ 0 \ \ 0 \ \ 1 \ \ 0 \ \ 0 \dots \dots]$$

a unit displacement in the vertical direction at node 3 in the direction of Y_3.

Carrying out the product $[u*]^T [X]$ in the equation above, we obtain just the external force component associated with the same degree of freedom, Y_3 i.e.,

$$Y_3 = [\delta*]^T[f]$$

---

17. We allow the system to be indeterminate as indicated in the figure.

We can cast this last equation into terms of member deformations (and member stiffness) and write:

$$Y_3 = [\delta^*]^T [k_{diag}][\delta]$$

And that is our special method for determining external forces of a statically determinate (or indeterminate) truss when all displacements are prescribed. It requires, first, solving compatibility for the "actual" member deformations $[\delta]$ given the "actual" prescribed displacements. We then solve *another* compatibility problem - one in which we apply a unit, or "dummy" displacement at the node we seek to determine an applied force component and in the direction of that force component. With the "dummy" member deformations determined from compatibility, we carry out the matrix multiplication of the last equation and there we have it.

We emphasize the difference between the two deformation vectors appearing in this equation; $[\delta]$ in plain font, is the vector of actual member deformations in the structure given the actual prescribed nodal displacements. $[\delta^*]$ starred, on the other hand, is some, originally arbitrary virtual deformation vector which satisfies compatibility - compatibility solution for member deformations corresponding to a unit displacement in the vertical direction at node 3.

We emphasize that our method does not require that we explicitly write out and solve the equilibrium equations for the system. We must, instead, compute compatible member deformations several times over.

A Generalization

We think of applying Displacement Method #1 at each degree of freedom in turn, and summarize all the relationships obtained for the required applied forces in one matrix equation. We do this by choosing

$$\left[\overset{*}{u}\right]^T = \begin{bmatrix} 1 & 0 & ..... & 0 & 0 \\ 0 & 1 & ..... & 0 & 0 \\ ..... & ..... & ..... & ..... & ..... \\ 0 & 0 & ..... & 1 & 0 \\ 0 & 0 & ..... & 0 & 1 \end{bmatrix} = [I] \qquad 12 \times 12$$

where each row represents a unit displacement in the direction of the "row[th]" degree of freedom.

The corresponding member deformations $[\delta^*]$ now takes the form of a 11 x 12 matrix whose "j[th]"column entries are the deformations engendered by the unit displacement of the "i^{th}" row above. (Note there are 11 members, hence 11 deformations and member forces).

We still have

$$[u*]^T [X] = [\delta*]^T [f] \qquad \text{where} \qquad [\delta*]^T = [u*]^T [A]$$

but now $[u*]^T$ is a 12 by 12 matrix, in fact the identity matrix. So we can write:

$$[X] = [\delta*]^T [f]$$

- an expression for all the required applied forces, noting that the matrix $[\delta*]^T$ is 12 by 11.

Some further manipulation takes us back to the matrix displacement analysis results of the last section. Eliminating $[\delta*]^T$ via the second equation on the line above, and setting $[u*]$ to the identity matrix, we have $\qquad [X] = [A][f]$

which we recognize as the equilibrium requirement. (But remember, in this world of prescribed displacements, analysts look upon this relationship as foreign; compatibility is their forte). We replace the real member forces in terms of the real member deformations, then, in turn, the real member deformations in terms of the prescribed and actual displacements and obtain

$$[X] = [A][k_{diag}][A]^T [u] \qquad \text{of} \qquad [X] = [K][u]$$

Design Exercise 5.1

You are a project manager for Bechtel with responsibility for the design and construction of a bridge to replace a decaying truss structure at the Alewife MBTA station in North Cambridge. Figure 1 shows a sketch of the current structure and Figure 2 a plan view of the site. The bridge, currently four lanes, is a major link in Route 2 which carries traffic in and out of Boston from the west. Because the bridge is in such bad shape, no three-axle trucks are allowed access. Despite its appearance, the bridge is part of a parkway system like Memorial Drive, Storrow Drive, et. al., meant to ring the city of Boston with greenery as well as macadam and concrete. In fact, the MDC, the Metropolitan District Commission, has a strong voice in the reconstruction project and they very much would like to stress the parkway dimension of the project. In this they must work with the DPW, the Department of Public Works. The DPW is the agency that must negotiate with the Federal Government for funds to help carry through the project. Other interested parties in the design are the immediate neighborhoods of Cambridge, Belmont, and Arlington; the environmental groups interested in preserving the neighboring wetlands. (Osprey and heron have been seen nearby.) Commuters, commercial interests – the area has experienced rapid development – are also to be considered.

1.1 Make a list of questions of things you might need to know in order to do your job.

1.2 Make a list of questions of things you might need to know to enable you to decide between proposing a four-lane bridge or a six-lane bridge.

1.3 Estimate the "worst-case" loads a four lane bridge might experience. Include "dead weight loading" as well as "live" loads.

1.4 With this loading:

a) sketch the shear-force and bending-moment diagram for a single span.

b) for a statically determinate truss design of your making, estimate the member loads by sectioning one bay, then another...

c) rough out the sizes of the major structural elements of your design.

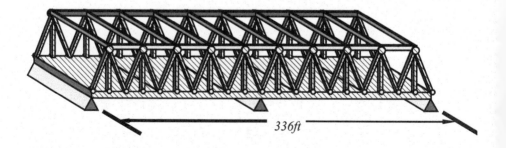

336ft

CONSERVATION COMMISSION FRUSTRATED AT ALEWIFE PLAN

(October 4, 1990, Belmont Citizen-Herald)   by Dixie Sipher Yonkers, Citizen-Herald correspondent[18]

Opponents of the planned $60 to $70 million Alewife Brook Parkway reconstruction can only hope the federal funding falls through or the state Legislature steps in at the eleventh hour with a new plan. Following a presentation by a Metropolitan District Commission planner on the Alewife Development proposal Tuesday night, the Belmont Conservation Commission expressed frustration over an approval process that appears to railroad a project of questionable benefit and uncertain impact, regardless of communities' concerns and requests. The Alewife project would widen Route 2 and redesign the truss bridge, access roads and access ramps on Route 2 near the Belmont-Arlington-Cambridge border. It also would extend Belmont's Brook Parkway significantly. Alewife Basin planner John Krajovick told the commission that MDC has grave concerns about the proposed transportation project and that, funding issues aside, it might be impossible to prevent the Massachusetts Department of Public Works' "preferred alternative" from being implemented. According to Krajovick, the MDC's concerns center around the loss of open space that will accompany the project, specifically the land along the eastern bank of Yates pond, the strip abutting the existing parkway between Concord Avenue and Route 2, the wetlands along the railroad right-of-way near the existing interim access road, and that surrounding the Jerry's pool site. "Our goal is to reclaim parkways to the original concept of them," said Krajovick. "It was Charles Elliot's vision to create a metropolitan park system – a kind of museum of unique open spaces...and use the parkways to connect them as linear parks." "The world has changed. They are no longer for pleasure vehicles only, but parkways, we feel, are a really important way to help to control growth and maintain neighborhood standards," he added. "We would like to see the character of this more similar to Memorial Drive and Storrow Drive as opposed to an expressway like Route 2." Krajovick outlined the MDC's further concerns with the project, citing its likely visual, physical, noise, and environmental impacts on surrounding neighborhoods. Projected to cost $60-$70 million, he said, the "preferred alternative" will also hurt a sensitive wetland area, the Alewife Reservation, in return for minimal traffic improvements. In spite of these concerns, Krajovick reported that the project is nearing a stage at which it becomes very difficult to prevent implementation. The Final Environmental impact Statement is expected to be submitted to the Federal Highway Department within a month. The same document will be used as the final report the state's Executive Office of Environmental Affairs. EOEA Secretary John DeVillars cannot stop the project once he receives that report. He can call for mitigating measures only. Krajovick noted a bill currently before the state Legislature's Transportation Committee could prohibit the project from going forward as presently designed. He took no position on that bill. Conservation Commission members, however, voiced doubts

---

18. Reprinted with permission of Harte-Hanks Community Newspapers, Waltham, MA

on the likelihood a passage in the face of the fiscal crisis and state elections that loom before legislators. In addition, Krajovick said that state budget cuts are expected to result in layoffs for nearly 600 of the MDC's staff of 1,000 workers, effectively decimating the agency. "Our hopes for a compromise solution may not happen," he said. Discouraged by Krajovick's dismal prognosis, Conservation Commission members expressed concern that there was nothing they could do to change the course of the project. The commission has been providing input on the project for 12 years with no results. In response to Krajovick's presentation, Commission member William Pisano called the need for updated impact studies, saying, "We agree with you. What we want to see is a lot more data and a more accurate realization of what we're playing ball with today." Commending the way in which concerned residents of Arlington, Cambridge and Belmont have gotten involved in the project, however, Krajovick said their thinking as a neighborhood rather than individual towns is a positive thing that has come from the project. Building on that team spirit, he said, the communities can raise their voice through formation of a Friends group and work toward the development of a master plan or restoration plan for the whole Alewife reservation area.

BRIDGE MEETING HIGHLIGHTS ISSUES

Belmont Citizen-Herald September 26, 1991   by Alin Kocharians, Citizen-Herald staff

Some 50 residents turned out out Tuesday night at Winn Brook School to hear a presentation by the state Metropolitan District Commission on the Alewife Brook Parkway Truss Bridge. MDC representatives previewed their Truss Bridge renovation and Parkway restoration plans. The Parkway segment affected is in Cambridge, between the Concord Avenue rotary and Rindge Avenue. Plans for the two-year project, MDC officials hope, will be completed by early 1992, with construction following in the spring of that year. Julia O'Brien, MDC's director of planning, said that the $12 million necessary for the project will be provided by the Legislature and federal grants. Once the bridge renovation is completed, the truck ban on it will be lifted, hopefully reducing truck traffic in Belmont. The renovation plans are 75 percent complete, according to John Krajovick, the MDC planner in charge of the project. The MDC is also visiting with Arlington and Cambridge residents, asking for input on the project's non-technical aspects. Residents and MDC representatives exchanged compliments in the first hour, but as the meeting wore on, the topics of cosmetic versus practical and local versus regional issues proved divisive. One Belmont resident summed up what appeared to be a common misgiving in town. "I don't want to cast stones, because it is a nice plan," said John Beaty of Pleasant Street, "but it doesn't solve the overall problem. I wish that I were seeing not just MDC here. There were two competing plans. It is the (State Department of Public Works') charter to solve the overall region's problem. I see those two as being in conflict." Beaty said that the DPW plan was presented two years ago to residents, when officials had said that the plan was 60 percent complete. Stanley Zdonik of Arlington agreed. "I am impressed with the MDC presentation, but what bothers me is, are you going to improve on the traffic flow?" he said. "You have got one bottleneck at one end, and another at the other." He said that the Concord Avenue and Route 2 rotaries at either end of the bridge should have traffic signals added, or be removed altogether. Krajovick replied that according to what the MDC's traffic engineer had told him, "historically, signalizing small rotaries actually backs up traffic even more." Belmont Traffic Advisory Committee member Marilyn Adams took issue with the decision not to add signals to the rotaries, and asked to see the study that produced this recommendation. Adams was also concerned with a "spill off" of traffic from the construction. "I can't guarantee people won't seek out other routes," including Belmont, O'Brien said. However, she added, she did not expect the impact to be very great, as the Parkway would still be open during construction. "We will make really a strong effort for a traffic mitigation" plan to be negotiated with the town, she said. Selectman Anne Paulsen also asked about the impact of traffic on the town. MDC representatives said that various traffic surveys were being conducted to find a way to relieve the traffic load on Belmont. Krajovick said that traffic problems in Belmont were regional questions, to be handled by local town officials, a point with which Paulsen disagreed. Paulsen said that she would prefer a more comprehensive plan for the region. Aside from the reconstruction of the Truss Bridge, she said, "I think the point of the people of Belmont is that...we want improvement in the roadway, so that we are relieved of some of the traffic." According to the plans, the new bridge will have four 11-foot lanes, one foot wider than the current width for each lane. There will also be a broader sidewalk, and many new trees planted both along the road and at the rotaries. There will be pedestrian passes over the road, and a median strip with greenery. The bridge will be made flat, so that motorists will have better visibility, engineering consultant Ray Oro said. It will be constructed in portions, so that two lanes will always be able to carry traffic, he said. According to Blair Hines of the landscaping firm of Halvorson Company, Inc., by the end of the project, "Alewife Brook Parkways will end up looking like Memorial Drive." All the talk about landscaping, Paulsen suggested with irony, "certainly calms the crowd."

5.4   Problems

5.1   If the springs are all of equal stiffness, k, the bar ABC rigid, and a couple M_0 is applied to the system, show that the forces in the springs are

$$F_A = -(5/7)M_0/H \qquad F_B = (1/7)M_0/H \qquad F_C = (4/7)M_0/H$$

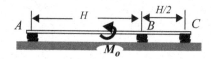

5.2   The problem show within the box was worked incorrectly by an MIT student on an exam. The student's work is shown immediately below the problem statement, again with the box.

i) Find and describe the error.

ii) Re-formulate the problem—that is, construct a set of equations from which you might obtain valid estimates for the forces in the two supporting members, *BD* and *CD*.

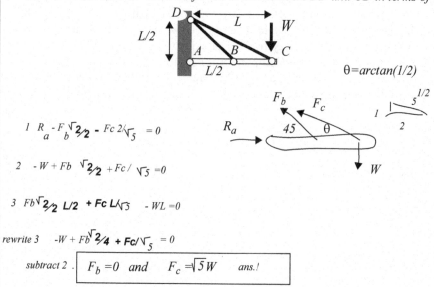

A **rigid** *beam is supported at the three pins, A,B, and C by the wall and the two elastic members of common material and identical cross-section. The rigid beam is weightless but carries an end load W. Find the forces in the members BD and CD in terms of W.*

5.3 A *rigid* beam is pinned supported at its left end and at midspan and the right end by two springs, each of stiffness k (force/displacement). The beam supports a weight P at mid span.

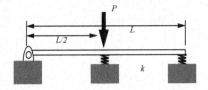

i) Construct a compatibility condition, relating the displacements of the springs to the rotation of the rigid beam.

ii) Draw an isolation and write out the consequences of force and moment equilibrium

iii) Using the force/deformation relations for the linear springs, express equilibrium in terms of the angle of rotation of the beam.

iv) Solve for the rotation, then for the forces of reaction at the three support points.

v) Sketch the shear force and bending moment diagram.

5.4 For the rigid stone block supported by three springs of Exercise 5.1, determine the displacements of and forces in the springs (in terms of W) if the spring at C is very, very stiff relative to the springs (of equal stiffness) at A and B.

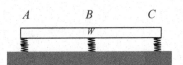

5.5 The stiffness matrix for the truss structure shown below left is

$$2\left[\frac{AE}{L}\right]\begin{bmatrix} cos^2 60 & 0 \\ 0 & sin^2 60 \end{bmatrix}\begin{bmatrix} u \\ v \end{bmatrix} = \begin{bmatrix} X \\ Y \end{bmatrix}$$

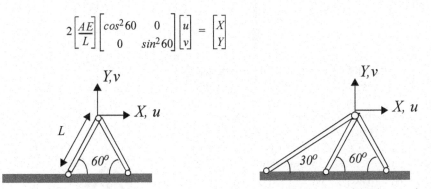

What if a third member, of the same material and cross-sectional area, is added to the structure to *stiffen it up*; how does the stiffness matrix change?

5.6   Without writing down any equations, *estimate* the maximum member tensile load within the truss structure shown below. Which member carries this load?

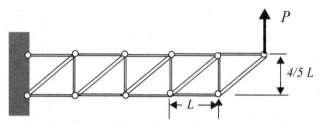

5.7   The truss show below is loaded at midspan with a weight P= 60 lbs. The member lengths and cross sectional areas are given in the figure. The members are all made of steel.

$$A_{top} = 0.01227 \ in^2$$

$$A_{diag} = 1.09 \ A_{top}$$

$$A_{bot} = 2.35 \ A_{top}$$

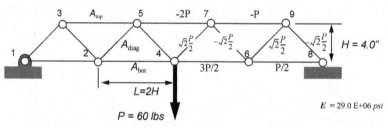

$$E = 29.0 \ E{+}06 \ psi$$

a) Verify that the forces in the members are as indicated.
b) Using Trussworks, determine the vertical deflections at nodes 2 and 4.

5.8   All members of the truss structure shown at the left are of the same material (Elastic modulus E), and have the same cross sectional area. Fill in the elements of the stiffness matrix.

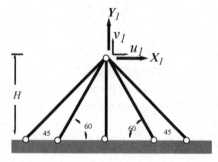

$$\left(\frac{AE}{H}\right) \begin{bmatrix} ? & ? \\ ? & ? \end{bmatrix} \begin{bmatrix} u_1 \\ v_1 \end{bmatrix} = \begin{bmatrix} X_1 \\ Y_1 \end{bmatrix}$$

5.9 For the three problems 1a, 1b, and 1c, state whether the problem posed is statically *determinate* or statically *indeterminate*. In this, assume all information regarding the geometry of the structure is given as well as values for the applied loads.

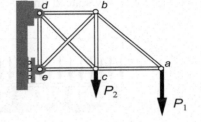

1a) Determine the force in member *ab*.

1b) Determine the force in member *bd*

1c) Determine the reactions at the wall.

5.10 The simple truss structure shown is subjected to a horizontal force P, directed to the right. The members are made of the same material, of Young's modulus E, and have the same cross-sectional area, A (for the first three questions).

i) Find the force acting in each of the two members *ab, bc*, in terms of P.

ii) Find the extension, (contraction), of each of the two members.

iii) Assuming small displacements and rotations, sketch the direction of the displacement vector of node *b*.

iv) Sketch the direction of the displacement vector if the cross-sectional area of *ab* is much greater than that of *bc*.

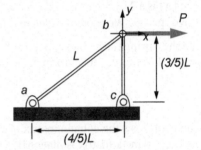

v) Sketch the direction of the displacement vector if the cross-sectional area of *ab* is much less than that of *bc*.

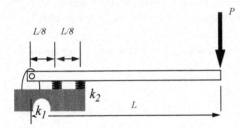

5.11 The rigid beam is pinned at the left end and supported also by two linear springs as shown.

What do the equilibrium requirements tell you about the forces in the spring and their relation to P and how they depend upon dimensions shown?

Assuming small deflections (let the beam rotate *cw* a small angle θ), what does compatibility of deformation tell you about the relationships among the contractions of the spring, the angle θ?

What do the constitutive equations tell you about the relations between the forces in the springs and their respective deflections?

Express the spring forces as a function of P if $k_2 = (1/4)k_1$

5.12   A rigid board carries a uniformly distributed weight, W/L. The board rests upon five, equally spaced linear springs, but each of a different stiffness.

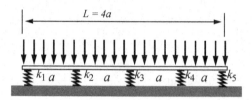

Show that the equations of equilibrium for the isolated, rigid board can be put in the form

$$\left[A\right] \cdot \left[F\right] = \begin{bmatrix} W \\ 0 \end{bmatrix}$$

where [A] is a 2 by 5 matrix and [F] is a 5 by 1 column matrix of the compressive forces in the five springs. Write out the elements of [A].

If the springs are linear, but each of a different stiffness, show that the matrix form of the force/deformation relations take the form

$$\left[F\right] = \left[k_{\text{diag}}\right] \cdot \left[\delta\right]$$

where the [δ] is a the column matrix of the spring deformations, taken as positive in compression, and the k matrix is diagonal.

Show that, if the beam is rigid and deformations are small then, in order for the spring deformations to be compatible, one with another, five equations must be satisfied (for small deformations). Letting u be the vertical displacement of the midpoint of the rigid beam - positive downward - and θ its counter-clockwise rotation, write out the elements of $[A]^T$ - the matrix relating the deformations of the spring to u and θ. Then show that the equations of equilibrium in terms of displacement take the form:

$$\left[A\right] \cdot \left[k_{\text{diag}}\right] \cdot \left[A\right]^T \cdot \begin{bmatrix} u \\ \theta \end{bmatrix} = \begin{bmatrix} W \\ 0 \end{bmatrix} \qquad \text{where } [A]^T \text{ is the transpose of } [A].$$

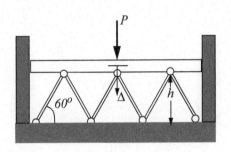

5.13 A rigid beam is constrained to move vertically without rotation. It is supported by a simple truss structure as shown in the figure. The truss members are made of aluminum ($E = 10.0 E+06 psi$; $= 70$ GPa). Their cross sectional area is $0.1 in^2 = 0.0645E-03 m^2$ The system bears a concentrated load P at midspan. The height of the platform above ground is $h = 36 in = 0.91 m$.

Let the vertical displacement be Δ. Determine the value of the stiffness of the system, the value for K in the relationship $P = K \Delta$

What is the vertical displacement if $P = 5,000$ lb $= 22,250$ N

What is the compressive stress in the members at this load?

5.14   A rigid beam rests on an elastic foundation. The distributed stiffness of the foundation is defined by the parameter β; the units of β are force per vertical displacement per length of beam. (If the beam were to displace downward a distance u_L without rotating, the total vertical force resisting this displacement would be just $\beta*u_L*L$). A heavy weight P rests atop the beam at a distance a to the right of center. The beam has negligible weight relative to P.

Letting the vertical displacement at the *left end* of the beam be u_L, and the rotation about this same point be θ, (clockwise positive), show that the requirements of force and moment equilibrium, applied to an isolation of the beam, give the following two equations for the displacement and rotation:

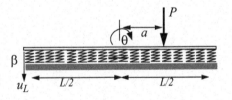

$$(\beta L) \cdot u_L + \left(\frac{\beta L^2}{2}\right) \cdot \theta = P$$

$$\left(\frac{\beta L^2}{2}\right) \cdot u_L + \left(\frac{\beta L^3}{3}\right) \cdot \theta = P \cdot (a + L/2)$$

Let $\Lambda = P/(\beta L^2)$, $\alpha = a/L$, and $z = u_L/L$ so to put these in non dimensional form. Then solve for the non-dimensional displacements z and θ. Explore the solution for special cases, e.g., $a = 0$, $-L/2$, $+L/2$. What form do the equilibrium equations take if you measure the vertical displacement at the center of the beam? (Let this displacement be u_o).

6

Strain

The study of the elastic behavior of statically determinate or indeterminate truss structures serves as a paradigm for the modeling and analysis of **all** structures in so far as it illustrates

- the isolation of a region of the structure prerequisite to imagining internal forces;
- the application of the equilibrium requirements relating the internal forces to one another and to the applied forces;
- the need to consider the displacements and deformations if the structure is redundant; [1]
- and how, if displacements and deformations are introduced, then the constitution of the material(s) out of which the structure is made must be known so that the internal forces can be related to the deformations.

We are going to move on, with these items in mind, to study the elastic behavior of shafts in torsion and of beams in bending with the aim of completing the task we started in an earlier chapter – among other objectives, to determine when they might fail. To prepare for this, we step back and dig in a bit deeper to develop more complete measures of deformation, ones that are capable of taking us beyond uniaxial extension or contraction. We then must relate these measures of deformation, the *components of strain at a point* to the *components of stress at a point* through some stress-strain equations. We address that task in the next chapter.

We will proceed without reference to truss members, beams, shafts in torsion, shells, membranes or whatever structural element might come to mind. We consider an arbitrarily shaped body, a *continuous solid body, a solid continuum*. We put on another special pair of eyeglasses, a pair that enables us to imagine what transpires *at a point* in a solid subjected to a load which causes it to deform and engenders strain along with some internal forces - the stresses of chapter 4. In our derivations that follow, we limit our attention to two dimensions: We first construct a set of strain measures in terms of the x,y (and z) components of displacement at a point. We then develop a set of stress/strain equations for a *linear, isotropic, homogenous, elastic solid*.

---

1. We of course must consider the deformations even of a determinate structure if we wish to estimate the displacements of points in the structure when loaded.

6.1   Strain: The Creature and its Components.

When a body displaces as a *rigid* body, points etched on the body will move through space but any arbitrarily chosen point will maintain the same distance from any other point–just a star in the sky maintains its position relative to all other stars, night after night, as the heavens rotate about the earth. Except, of course, for certain "wandering stars" which do not maintain fixed distances among them-selves or from the others.

But when a body *deforms*, points move relative to one another and distances between points change. For example, when the bar shown below is pulled with an end load P along its axis we know that a point at the end will displace to the right, say a distance u_L, relative to a point at the fixed, left end of the bar.

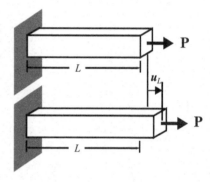

Assuming the bar is *homogenous*, that is, its constitution does not change as we move in from the end of the bar, we anticipate that the displacement relative to the fixed end will decrease. At the wall it must be zero; at the mid point we might anticipate it will be $u_L/2$. Indeed, this was the essence of our story about the behavior of an elastic rod in a uniaxial tension test.

There we had $P = (AE/L) \cdot \delta = k \cdot \delta$

The stiffness k is inversely proportional to the length of the rod so that, if the *same* end load is applied to bars of different lengths, the displacement of the ends will be proportional to their lengths, and the ratio of δ to L will be constant.

In our mind, then, we can imagine the horizontal rod shown above cut through at its midpoint. As far as the remaining, left portion is concerned, it is fixed at its left end and sees an a load P at its right end. Now since it has but half the length, its *end* will displace to the right but $u_L/2$.

We can continue this thought experiment from now to eternity; each time we make a cut we will obtain a midpoint displacement which is one-half the displace-ment at the right end of the previously imagined section. This of course assumes the bar is uniform in its cross-sectional area and material properties —that is, the bar is *homogeneous*. We summarize this result neatly by writing

$$u(x) = (u_L/L) \cdot x$$

where the factor, (u_L/L), is a measure of the extensional strain of the bar, defined as the ratio of the change in length of the bar to its original length.

This brief thought experiment gives us a way to define a measure of exten-sional strain *at a point*. We say, at any point in the bar, that is, at any x,

$$\varepsilon_x = \lim_{\Delta x \to 0} (\Delta u/\Delta x) \Rightarrow \varepsilon_x = \frac{\partial u}{\partial x}$$

For the homogenous bar under end load P we see that ε_x is a constant; it does not vary with x. We might claim that the end displacement, u_L is *uniformly distributed* over the length; that is, the relative displacement of any two points, equidistant apart in the undeformed state, is a constant; but this is not the usual way of speaking nor, other than for a truss element, is it usually the case.

The partial derivative implies that, u, the displacement component in the x direction, can be a function of spatial dimensions other than x alone; that is, for an arbitrary solid, with things changing as one moves in any of the three coordinate directions, we would have $u = u(x,y,z)$. We turn to this more general situation now.

Exercise 6.1

What do I need to know about the displacements of points in a solid in order to compute the extensional strain at the point P, arbitrarily taken, in the direction of t_0, also arbitrarily chosen, as the body deforms from the state indicated at the left to that at the right?

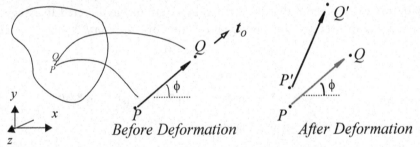

Before Deformation   *After Deformation*

We designate the extensional strain at P in the direction of t_0 by ε_{PQ}. Our task is to see what we need to know in order to evaluate the limit

$$\varepsilon_{PQ} = \lim_{PQ \to 0} (P'Q' - PQ)/(PQ)$$

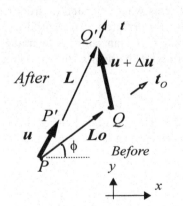

To do this, we draw another picture of the undeformed and deformed differential line element, PQ. together with the displacements of its endpoints. Point P's displacement to P' is shown as the vector, *u,* while the displacement of point Q, some small distance away, is designated by $u+\Delta u$.

This now looks very much like the representation used in the last chapter to illustrate and construct an expression for the extension of a truss member as a function of the horizontal and vertical components of displacement at its two ends. That's why I have introduced the vectors L_0, and L for the *directed line segments* PQ, $P'Q'$ respectively though they are in fact meant to be small, differential lengths. Proceeding in the same way as we did in our study of the truss, we write, as a consequence of vector addition,

$$\mathbf{u} + \mathbf{L} = \mathbf{u} + \Delta\mathbf{u} + \mathbf{L_0}$$

which yields an expression for $\Delta \mathbf{u}$ in terms of the vector difference of the two directed line segments, namely $\Delta\mathbf{u} = \mathbf{L} - \mathbf{L_0}$

We now introduce a most significant constraint. We assume, as we did with the truss, that *displacements and rotations are small* – displacements relative to some characteristic length of the solid, rotations relative to a radian. This should not be read as implying our analysis is of limited use. Most structures behave, i.e., deform, according to this constraint and, as we have seen in our study of a truss structure, it is entirely consistent with our writing the *equilibrium equations with respect to the undeformed configuration*. In fact *not* to do so would be erroneous.

Explicitly this means we will take

$$\mathbf{t} \approx \mathbf{t_0} \qquad \text{so that} \qquad |\mathbf{L}| = \mathbf{t} \bullet \mathbf{L} \approx \mathbf{t_0} \bullet \mathbf{L}$$

With this we can claim that the change in length of the directed line segment, PQ, in moving to $P'Q'$, is given by the projection of $\Delta\mathbf{u}$ upon PQ that is, since

$$P'Q' - PQ = |\mathbf{L}| - |\mathbf{L_0}|$$

we have

$$P'Q' - PQ = \mathbf{t_0} \bullet \mathbf{L} - \mathbf{t_0} \bullet \mathbf{L_0} = \mathbf{t_0} \bullet (\mathbf{L} - \mathbf{L_0}) = \mathbf{t_0} \bullet \Delta\mathbf{u}$$

where $\mathbf{t_0}$ is, as before, a unit vector in the direction of PQ.

$$\mathbf{t_0} = \cos\phi \cdot \mathbf{i} + \sin\phi \cdot \mathbf{j}$$

From here on in, constructing an expression for ε_{PQ} requires the machine-like evaluation of the scalar product, $\mathbf{t_0} * \Delta\mathbf{u}$, the introduction of the partial derivatives of the scalar components of the displacement taken with respect to position, and the manipulation of all of this into a form which reveals what's needed in order to compute the relative change in length of the arbitrarily oriented, differential line segment, PQ. We work with respect to a rectangular cartesian coordinate frame, x,y, and define the horizontal and vertical components of the displacement vector $\mathbf{u}$ to be u,v respectively[2]. That is, we set

$$\mathbf{u} = u(x, y) \cdot \mathbf{i} + v(x, y)\mathbf{j}$$

where the coordinates x,y label the position of the point P. The differential change in the displacement vector in moving from P to Q, a small distance which in the limit will go to zero, may then be written

---

2. In the following be careful to distinguish between the scalar u and the vector **u**; the former is the x component of the latter.

$$\Delta \mathbf{u} = \Delta u(x, y) \cdot \mathbf{i} + \Delta v(x, y) \cdot \mathbf{j}$$

Carrying out the scalar product, we obtain for the change in length of PQ:

$$P'Q' - PQ = \mathbf{t}_0 \bullet \Delta \mathbf{u} = (\Delta u) \cdot cos\phi + (\Delta v) \cdot sin\phi$$

We next approximate the small changes in the horizontal and vertical scalar components of displacement by the products of their slopes at P taken with the appropriate differential lengths along the x and y axes as we move to point Q. That

is[3] $\Delta u(x, y) \approx \left(\dfrac{\partial u}{\partial x}\right)\Delta x + \left(\dfrac{\partial u}{\partial y}\right)\Delta y$ and $\Delta v(x, y) \approx \left(\dfrac{\partial v}{\partial x}\right)\Delta x + \left(\dfrac{\partial v}{\partial y}\right)\Delta y$

We have then

$$(P'Q' - PQ)/(PQ) \approx \left[\left(\frac{\partial u}{\partial x}\right)\Delta x + \left(\frac{\partial u}{\partial y}\right)\Delta y\right](cos\phi / L_o) + \left[\left(\frac{\partial v}{\partial x}\right)\Delta x + \left(\frac{\partial v}{\partial y}\right)\Delta y\right](sin\phi / L_o)$$

where I have introduced L_o for the original length PQ.

This is an approximate relationship because the changes in the horizontal and vertical components of displacement are only approximately represented by the first partial derivatives. *In the limit*, however, as the distance PQ, and hence as Δx, Δy approaches zero, the approximation may be made as accurate as we like. Note also, that the ratios $\Delta x/L_o$, $\Delta y/L_o$ approach $cos\phi$ and $sin\phi$ respectively.

We obtain, finally, letting PQ go to zero, the following expression for the extensional strain *at the point P* in the direction PQ:

$$\varepsilon_{PQ} = \left(\frac{\partial u}{\partial x}\right) cos^2\phi + \left(\frac{\partial u}{\partial y} + \frac{\partial v}{\partial x}\right) cos\phi\, sin\phi + \left(\frac{\partial v}{\partial y}\right) sin^2\phi$$

It appears that in order to compute ε_{PQ} in the direction ϕ we *need to know* the four first partial derivatives of the scalar components of the displacement at the point P. In fact, however, we do not need to know all four partial derivatives since it is enough to know the **three** bracketed terms appearing above. Think of computing ε_{PQ} for different values of ϕ; knowing the values for the three bracketed terms will enable you to do this.

The relationship above is a very important piece of machinery. It tells us how to compute the extensional strain in any direction, defined by ϕ, at any point, defined by x,y, in a body. In what follows, we call the three quantities within the brackets the *three scalar components of strain at a point.* But first observe:

- If we set ϕ equal to zero in the above, which is equivalent to setting PQ out along the x axis, we obtain, as we would expect, that $\varepsilon_{PQ} = \varepsilon_x$, the extensional strain at P in the x direction, i.e.,

$$\varepsilon_x = \left(\frac{\partial u}{\partial x}\right)$$

---

3. It is easy to be confused in the midst of all these partial derivatives. It's worth taking five minutes to try to sort them out.

- Our machinery is thus consistent with our previous definition of ε_x for uniaxial loading of a bar fixed at one end and lying along the x axis.

- If, in the same way, we set ϕ equal to a right angle, we obtain

$$\varepsilon_{PQ} = \left(\frac{\partial v}{\partial y}\right)$$

which can be read as the extensional strain at P in the direction of a line segment along the y axis. We call this ε_y. That is

$$\varepsilon_y = \left(\frac{\partial v}{\partial y}\right)$$

- The meaning of the term $\frac{\partial u}{\partial y} + \frac{\partial v}{\partial x}$ is best extracted from a sketch; below we show how the term $\frac{\partial v}{\partial x}$ can be interpreted as the angle of rotation, about the z axis, of a line segment PQ along the x axis. For small rotations we can claim

$$\alpha \sim \tan\alpha = \frac{\left(\frac{\partial v}{\partial x}\right)\Delta x}{\Delta x}$$

Similarly, the term $\delta u/\delta y$ can be interpreted as the angle of rotation of a line segment along the y axis, but now, if positive, about the *negative* z axis. The figure below shows the meaning of both terms.

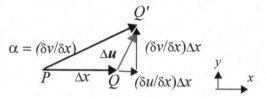

The sum of the two terms is the change in the right angle, PQR at point P. If it is a positive quantity, the right angle of the first quadrant has decreased. We define this sum to be a *shear strain* component at point P and label it with the symbol γ_{xy}.

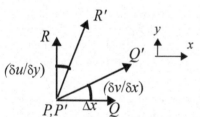

- Building on the last figure, we define a *rotation* at the point P as the **average** of the rotations of the two x,y, line segments. That is, we define

$$\omega_{xy} = (1/2)\left(\frac{\partial v}{\partial x} - \frac{\partial u}{\partial y}\right)$$

Note the negative sign to account for the different directions of the two line segment rotations. If, for example, $\delta v/\delta x$ is positive, and $\delta u/\delta y = -\delta v/\delta x$ then there is *no shear strain*, no change in the right angle, but there *is* a *rotation*, of magnitude $\delta v/\delta x$ positive about the z axis at the point P. These three quantities $\varepsilon_x, \gamma_{xy}, \varepsilon_y$ are the three components of strain at a point.

$$\varepsilon_x \equiv \frac{\partial u}{\partial x} \qquad \gamma_{xy} \equiv \left(\frac{\partial v}{\partial x} + \frac{\partial u}{\partial y}\right) \qquad \varepsilon_y \equiv \frac{\partial v}{\partial y}$$

If we know the way $\varepsilon_x(x,y)$, $\gamma_{xy}(x,y)$, and $\varepsilon_y(x,y)$ vary, we say we know the *state of strain* at any point in the body. We can then write our equation for computing the extensional strain in any arbitrary direction in terms of these three strain components associated with the x,y frame at a point as:

$$\varepsilon_{PQ} = \varepsilon_x \cdot \cos^2\phi + \gamma_{xy} \cdot \cos\phi\sin\phi + \varepsilon_y \cdot \sin^2\phi$$

Finally, note that if we are given the displacement components as continuous functions x *and* y we can, by taking the appropriate partial derivatives, compute a set of strain functions, also continuous in x,y. On the other hand, going the other way, given the three strain components, $\varepsilon_x, \gamma_{xy}, \varepsilon_y$ as continuous functions of position, we cannot be assured that we can determine unique, continuous functions for the two displacement components from an integration of the strain-displacement relations. We say that the strains represent a *compatible* state of deformation only if we can do so, that is, only if we can construct a continuous *displacement field* from the strain components.

Exercise 6.2

For the planar displacement field defined by

$$u(x, y) = -\kappa \cdot xy \qquad\qquad v(x, y) = \kappa \cdot x^2/2$$

where $\kappa = 0.25$, *sketch the locus of the edges of a 2x2 square, centered at the origin, after deformation and construct expressions for the strain components* $\varepsilon_x, \varepsilon_y$ *and* γ_{xy}

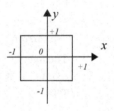

We start by evaluating the components of strain; we obtain

$$\varepsilon_x \equiv \frac{\partial u}{\partial x} = -\kappa y \qquad \gamma_{xy} \equiv \left(\frac{\partial v}{\partial x} + \frac{\partial u}{\partial y}\right) = -\kappa x + \kappa x = 0 \qquad \varepsilon_y \equiv \frac{\partial v}{\partial y} = 0$$

We see that the only non zero strain is the extensional strain in the x direction at every point in the plane. In particular, right angles formed by the intersection of a line segment in the x direction with another in the y direction remain right angles since the shear strain vanishes. The average rotation of these intersecting line segments at each and every point is found to be

$$\omega_{xy} = (1/2)\left(\frac{\partial v}{\partial x} - \frac{\partial u}{\partial y}\right) = \kappa x$$

We sketch the locus of selected points and line segments below:

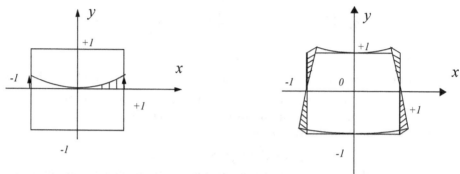

Focus first, on the figure at the left above which shows the deformed position of the points that originally lay along the *x* axis, at *y*=0. The vertical component of displacement *v* describes a parabola in the deformed state. Furthermore, the points along the *x* axis experience no horizontal displacement.

On the other hand, the points off the *x* or the *y* axis all have a horizontal component of displacement - as well as vertical. Consider now the figure above right. For example the point (1,1) moves to the left a distance *0.25* while moving up a distance *0.125*. Below the *x* axis, however, the point originally at (1,-1) moves to the right *0.25* while it still displaces upward the same *0.125*. The shaded lines are meant to indicate the *u* at each point.

Observe

- The state of strain does not vary with *x*, but does so with *y*.

- Right angles formed by *x-y* line segments remain right angles, that is, the *shear strain* is zero.

- The average rotation of these right angles *does* vary with *x* but not with *y*. Note too that we have seemingly violated the assumption of small rotations. We did so in order to better illustrate the deformed pattern.

6.2   Transformation of Components of Strain

The axial stress in a truss member is related to the extensional strain in the member through an equation that looks very much like that which relates the force in a spring to its deflection. We shall relate all stress and strain components through some more general constitutive relations — equations which bring the specific properties of the material into the picture. But stress and strain are "relations" in another sense, in a more abstract, mathematical way: They are both the **same kind of mathematical entity**. The criterion and basis for this claim is the following: **The components of stress and strain at a point transform according to the**

same equations. By transform we mean change; by change we mean change due to a rotation of our reference axis at the point.

Our study of how the components of strain and stress transform is motivated as much by the usefulness of this knowledge in engineering practice as by visions of mathematical elegance and sophistication[4]. For, although this section could have been labeled *the transformation of symmetric, second-order tensors*, we have already seen an example, back in our study of stress, an example suggesting the potential utility of the component transformation machinery. We do an exercise very similar to that we tackled before to refresh our memory.

Exercise 6.3

*Three strain gages, attached to the surface of a solid shaft in torsion in the directions **x, y**, and x' measure the three extensional strains*

$$\varepsilon_x = 0 \qquad \varepsilon_y = 0 \qquad \text{and} \qquad \varepsilon_{x'} = 0.00032$$

Estimate the shear strain γ_{xy}.

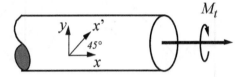

Let's work backwards. No one says you have to work forward from the "givens" straight through to the answer[5].

We are given the values of three extensional strains measured at a point on the surface of the shaft[6]. The task is to determine the shear strain at the point from the three, measured extensional strains.

From the previous section we know that the extensional strain in the x' direction - thinking of that direction as "PQ" - can be expressed as

$$\varepsilon_{PQ} = \varepsilon_x \cdot \cos\phi^2 + \gamma_{xy} \cdot \cos\phi \sin\phi + \varepsilon_y \cdot \sin\phi^2,$$

which tells me how to compute the extensional strain in some arbitrarily oriented direction at a point, as defined by the angle ϕ, given the state of strain at the point as defined by the three components of strain with respect to an x,y axis.

Working backwards, I will use this to compute the shear strain γ_{xy} given knowledge of the extensional strain ε_{PQ} where PQ is read as the direction of the gage x'

---

4. Katie: See Reid...I told you so!
5. This is characteristic of most work, not only in engineering but in science as well. The desired end state – the answer to the problem, the basic form of a design, the theorem to be proven, the character of the data to be collected – is usually known at the outset. There are really very few surprises in science or engineering in this respect. What **is** surprising, and exciting, and rewarding is that you can manage to construct things to come out right and they work according to your expectations.
6. It's not really a point but a region about the size of a small coin.

oriented at 45° to the axis of the shaft and pasted to its surface. Now both ε_x and ε_y are zero[7] so this equation gives

$$0.000032 = \gamma_{xy} \cdot (1/2) \qquad \text{or} \qquad \gamma_{xy} = 0.00064$$

Observe:

- If the strains ε_x and ε_y were different from zero we would still use this relationship to obtain an estimate of the shear strain. The former would provide us with direct estimates of any axial or hoop strain.

- I can graphically interpret this equation for determining the shear strain by constructing a compatible (continuous) displacement field from the strain components ε_x, ε_y, and γ_{xy}. Note this is not the only displacement field I might generate that is consistent with these strain components but it will serve to illustrate the relationship.

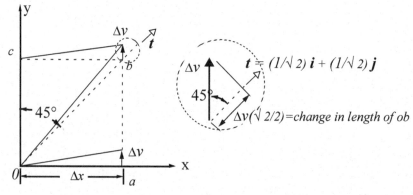

With the shaft oriented horizontally and twisted as shown, I take the displacement component, $u(x,y)$ to be zero and $v(x,y)$ to be proportional to x but independent of y. Then the points a and b both displace vertically a distance Δv with respect to points 0 and c. The extension of the diagonal $0b$ is, for small displacements and rotations, the projection of Δv at b upon the diagonal itself. So the change in length is given by $\Delta v /(\sqrt{2})$. Its original length is $\sqrt{2} \cdot \Delta x$ so we can write $\varepsilon_{x'} = \varepsilon_{ob} = \dfrac{(\Delta v /\Delta x)}{2}$

But, again for small rotations, $\Delta v/\Delta x = \gamma_{xy}$ the decrease in the right angle, the shear strain. Thus, as before,

$$\varepsilon_{x'} = \gamma_{xy} \cdot (1/2)$$

---

7. More realistic values would be some small, insignificant numbers due to *noise* or slight imbalance in the apparatus used to measure, condition, and amplify the signal produced by the strain gage. Even so, if the shaft was subject to forms of loading other than, and in addition to the torque we seek to estimate, and these engendered significant strains in the a and c directions we would still make use of this relationship in estimating the shear strain.

This exercise illustrates an application of the rules governing the transformation of the components of strain at a point. That's now the way we read the equation we saw in the previous section – as a way to obtain the **extensional strain** along one axis of an arbitrarily oriented coordinate frame at a point in terms of the strain components known with respect to some reference coordinate frame.

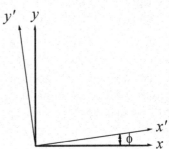

For example, if I let the arbitrarily oriented frame be labeled x'-y', then the **extensional** strain components relative to this new axis system can be written in terms of the strain components associated with the original, x-y frame as

$$\varepsilon'_x = \varepsilon_x \cdot cos\phi^2 + \gamma_{xy} \cdot cos\phi sin\phi + \varepsilon_y \cdot sin\phi^2$$

$$\varepsilon'_y = \varepsilon_x \cdot cos\phi^2 - \gamma_{xy} \cdot cos\phi sin\phi + \varepsilon_y \cdot sin\phi^2$$

In obtaining the expression for the extensional strain in the y' direction, I substituted $\phi + \pi/2$ for ϕ in the first equation.

But there is more to the story. I must construct an equation that allows me to compute the **shear strain**, γ'_{xy} relative to the arbitrarily oriented frame, x'y'. To do so I make use of the same graphical methods of the previous section.

The figure below left shows the orientation of my reference x-y axis and the orientation of an arbitrarily oriented frame x'-y'. PQ is a differential line element in the undeformed state lying along the x' axis. t is a unit vector along PQ; e is a unit vector perpendicular to PQ in the sense shown. Δx, Δy are the horizontal and vertical coordinates of Q relative to the origin of the reference frame.

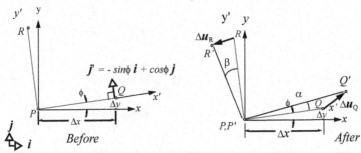

Before   *After*

On the right we show the position of PQ in the deformed state as $P'Q'$. The displacement of point Q *relative to* P is shown as Δu_Q. The angle α is the (small) rotation of the line element PQ. This is what we seek to express in terms of the strain components ε_x, ε_y *and* γ_{xy} at the point. We will also determine the rotation of a line element along the y' axis. Knowing these we can compute the change in the right angle QPR, the shear strain component with respect to the x'-y' system which we will mark with a "prime", γ_{xy}'.

The angle α is given approximately by $\alpha = \Delta u \bullet j'/(PQ)$ where j' is perpendicular to PQ.

The displacement vector we write as $\Delta u = \Delta u \cdot i + \Delta v \cdot j$ which, to first order may be written in terms of the partial derivatives of the scalar components of the relative displacement of Q.

$$\Delta u = \left(\Delta x \frac{\partial u}{\partial x} + \Delta y \frac{\partial u}{\partial y}\right) i + \left(\Delta x \frac{\partial v}{\partial x} + \Delta y \frac{\partial v}{\partial y}\right) j$$

and the unit vector is $j' = -\sin\phi \cdot i + \cos\phi \cdot j$

Carrying out the scalar, dot product, noting that
$$\Delta y/PQ = \cos\phi \qquad \text{and} \qquad \Delta y/PQ = \sin\phi$$

we obtain

$$\alpha \approx -\sin\phi\left(\cos\phi \frac{\partial u}{\partial x} + \sin\phi \frac{\partial u}{\partial y}\right) + \cos\phi\left(\cos\phi \frac{\partial v}{\partial x} + \sin\phi \frac{\partial v}{\partial y}\right)$$

Or collecting terms

$$\alpha \approx \sin\phi \cos\phi\left(\frac{\partial v}{\partial y} - \frac{\partial u}{\partial x}\right) + \cos^2\phi \frac{\partial v}{\partial x} - \sin^2\phi \frac{\partial u}{\partial y}$$

I obtain the angle β the rotation of a line segment PR originally oriented along the y' axis most simply by letting ϕ go to $\phi + \pi/2$ in the above equation for the angle α. Thus

$$\beta \approx -\sin\phi \cos\phi\left(\frac{\partial v}{\partial y} - \frac{\partial u}{\partial x}\right) + \sin^2\phi \frac{\partial v}{\partial x} - \cos^2\phi \frac{\partial u}{\partial y}$$

the diminution in the right angle QPR is just $\alpha - \beta$ so I obtain:

$$\gamma'_{xy} \approx 2\sin\phi \cos\phi\left(\frac{\partial v}{\partial y} - \frac{\partial u}{\partial x}\right) + (\cos^2\phi - \sin^2\phi)\left(\frac{\partial v}{\partial x} \frac{\partial u}{\partial y}\right)$$

which, in terms of the strain components associated with the x,y axes becomes

$$\gamma'_{xy} = 2(\varepsilon_y - \varepsilon_x) \cdot \sin\phi \cos\phi + \gamma_{xy} \cdot (\cos\phi^2 - \sin\phi^2)$$

With this I have all the machinery I need to compute the components of strain with respect to one orientation of axes at a point **given** their values with respect to another. I summarize below, making use of the *double angle* identities for the $\cos\phi$ and the $\sin\phi$, namely, $\cos 2\phi = \cos^2\phi - \sin^2\phi$ and $\sin 2\phi = 2\sin\phi\cos\phi$.

$$\varepsilon'_x = \left[\frac{(\varepsilon_x + \varepsilon_y)}{2}\right] + \left[\frac{(\varepsilon_x - \varepsilon_y)}{2}\right] \cdot \cos 2\phi + (\gamma_{xy}/2)\sin 2\phi$$

$$\varepsilon'_y = \left[\frac{(\varepsilon_x + \varepsilon_y)}{2}\right] - \left[\frac{(\varepsilon_x - \varepsilon_y)}{2}\right] \cdot \cos 2\phi - (\gamma_{xy}/2)\sin 2\phi$$

$$(\gamma'_{xy}/2) = -\left[\frac{(\varepsilon_x - \varepsilon_y)}{2}\right] \cdot \sin 2\phi + (\gamma_{xy}/2)\cos 2\phi$$

I have introduced a common factor of (1/2) in the equation for the shear strain for the following reasons: If you compare these transformation relationships with those we derived for the components of stress, back in chapter 4, you will see they are identical in form if we identify the normal strain components with their corresponding normal stress components but we must identify τ_{xy} with $\gamma_{xy}/2$.

One additional relationship about deformation follows from our analysis: If I average the angular rotations of the two orthogonal line segments *PQ* and *PR*, I obtain an expression for what we define as the *rotation* of the *x'-y'* axes at the point. This produces

$$\omega'_{xy} = \left(\frac{1}{2}\right)(\alpha + \beta) = \left(\frac{1}{2}\right)\left(\frac{\partial v}{\partial x} - \frac{\partial u}{\partial y}\right) = \omega_{xy}$$

This, we note, is identical to ω_{xy} which is what justifies labeling this measure of deformation a *rigid body rotation*. It is also *invariant* of the transformation; regardless of the orientation of the coordinate frame at the point, you will always get the same number for this measure of rotation.

Exercise 6.4

A "bug" in my graphics software distorts the image appearing on my monitor. Horizontal lines are stretched 1%; vertical lines are compressed 5% and there is a distortion of the right angles formed by the intersection of horizontal and vertical lines of approximately 3° – a decrease in right angle in the first quadrant. Estimate the maximum extensional distortion I can anticipate for an arbitrarily oriented line drawn by my software. What is the orientation of this particular line relative to the horizontal?

I seek a maximum value for the extensional strain at a point — the extensional strain of an arbitrarily oriented line segment which is maximum. Any point on the screen will serve; we are working with a *homogeneous* state of strain, one which does not vary with position. I also of course want to know the direction of this line segment. The equation above for ε_x' shows the extensional strain as a function of

ϕ; we differentiate with respect to ϕ seeking the value for the angle which will give a maximum (or minimum) extensional strain. I have:

$$\frac{d\varepsilon'_x}{d\phi} = -(\varepsilon_y - \varepsilon_x)sin2\phi + \gamma_{xy} \cdot cos2\phi = 0$$

which I manipulate to

$$tan2\phi = \gamma_{xy}/(\varepsilon_y - \varepsilon_x)$$

Now the three x,y components of strain are $\varepsilon_x = 0.01$, $\varepsilon_y = -0.05$, *and* $\gamma_{xy} = 3/57.3 = 0.052$. The above relationship, because of the behavior of the tangent function, will give me two roots within the range $0 < \phi < 360^o$, hence two values of ϕ.

I obtain two possibilities for the angle of orientation of maximum (or minimum) extensional strain, $\phi = 20.6^o$ *and* $\phi = 20.6 + 90^o = 110.6^o$ One of these will correspond to a maximum extensional strain, the other to a minimum. Note that we can read the second root as an extensional strain in a direction perpendicular to that associated with the first root. In other words, if we evaluate both ε_x' and ε_y' for a rotation of $\phi = 20.6^o$ we will find one a maximum the other a minimum. This we do now.

Taking then, $\phi = 20.6^o$ I obtain for the extensional strain in that direction,
$\varepsilon_I = 0.0197$

about two percent extension. The extensional strain at right angles to this I obtain from the equation for ε_y', a strain along an axis 110.6^o around from the horizontal, $\varepsilon_{II} = -0.0597$,

about six percent contraction. This latter is the maximum extensional distortion, a contraction of 5.97%. We illustrate the situation below.

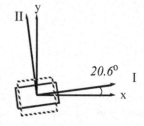

Observe

- We call this pair of extreme values of extensional strain at a point, one a maximum, the other a minimum, *the principal strains*; the axes they are associated with are called *the principal axes*.

- The **shear strain associated with the principal axes is zero, always**. This follows from comparing the equation we derived by setting the derivative of the arbitrarily oriented extensional strain with respect to angle of rotation, namely

$$tan(2\phi) = \gamma_{xy}/(\varepsilon_x - \varepsilon_y)$$

with the equation for the transformed component γ_{xy}'. If the former is satisfied then the shear must vanish.

6.3 Mohr's Circle

Our working up of the transformation relations for stress and for strain and our exploration of their meaning in terms of extreme values has required considerable mathematical manipulation. We turn again to our graphical rendering of these relationships called *Mohr's Circle*. I have set out the rules for constructing the circle for a particular state of stress. What I seek now is to show the "sameness" of the transformation relations for strain components.

First, I repeat the transformation equations for a two-dimensional state of stress.

$$\sigma'_x = \left[\frac{(\sigma_x + \sigma_y)}{2}\right] + \left[\frac{(\sigma_x - \sigma_y)}{2}\right] \cdot \cos 2\phi + \sigma_{xy}\sin 2\phi$$

$$\sigma'_y = \left[\frac{(\sigma_x + \sigma_y)}{2}\right] - \left[\frac{(\sigma_x - \sigma_y)}{2}\right] \cdot \cos 2\phi - \sigma_{xy}\sin 2\phi$$

$$\sigma'_{xy} = -\left[\frac{(\sigma_x - \sigma_y)}{2}\right] \cdot \sin 2\phi + \sigma_{xy}\cos 2\phi$$

and now the transformation equations for a two-dimensional state of strain:

$$\varepsilon'_x = \left[\frac{(\varepsilon_x + \varepsilon_y)}{2}\right] + \left[\frac{(\varepsilon_x - \varepsilon_y)}{2}\right] \cdot \cos 2\phi + (\gamma_{xy}/2)\sin 2\phi$$

$$\varepsilon'_y = \left[\frac{(\varepsilon_x + \varepsilon_y)}{2}\right] - \left[\frac{(\varepsilon_x - \varepsilon_y)}{2}\right] \cdot \cos 2\phi - (\gamma_{xy}/2)\sin 2\phi$$

$$(\gamma'_{xy}/2) = -\left[\frac{(\varepsilon_x - \varepsilon_y)}{2}\right] \cdot \sin 2\phi + (\gamma_{xy}/2)\cos 2\phi$$

Comparing the two sets, we see they are the same **if we compare half the shear strain with the corresponding shear stress.** This means we can use the same Mohr's Circle for stress when doing strain transformation problems. All we need do is think of the vertical axis as being a measure of $\gamma/2$.

6.4   Problems

6.1   *Show that* for the thin circular hoop subject to an axi-symmetric, radial extension u_r, that the circumferential extensional strain, can be expressed as

$$\varepsilon_\theta = (L - L_o)/L_o = u_r/r_o$$

where L_O is the original, undeformed circumference.

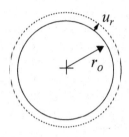

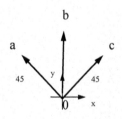

6.2   Three strain gages are mounted in the directions shown on the surface of a thin plate. The values of the extensional strain each measures is also shown in the figure.
i) Determine the shear strain component γ_{xy} at the point with respect to the xy axes shown.
ii) What orientation of axes gives extreme values for the extensional strain components at the point.
iii) What are these values?

6.3   Three strain gages measure the extensional strain in the three directions 0a, 0b and 0c at "the point 0". Using the relationship we derived in class

$$\varepsilon_{PQ} = \varepsilon_x \cos^2\phi + \gamma_{xy}\cos\phi\sin\phi + \varepsilon_y \sin^2\phi$$

find the components of strain with respect to the xy axis in terms of ε_a, ε_b and ε_c

6.4 A strain gage rosette, fixed to a flat, thin plate, measures the following extensional strains

ε_a = 1. E-04
ε_b = 1. E-04
ε_c = 2. E-04

Determine the state of strain at the point, expressed in terms of components relative to the xy coordinate frame shown.

6.5   A two dimensional displacement field is defined by

$$u(x,y) = \frac{-\alpha}{2} \cdot y \qquad \text{and} \qquad v(x,y) = \frac{\alpha}{2} \cdot x$$

Sketch the position of the
points originally lying along
the *x* axis, the line *y = 0*, due to
this displacement field.
Assume α is very much less
than 1.0.

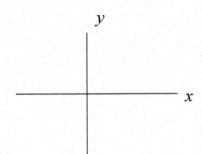

Likewise, on the same sketch,
show the position of the points
originally lying along the *y*
axis, the line *x = 0*, due to this
displacement field.

Likewise, on the same sketch, show the position of the points originally lying along the
line *y=x*, due to this displacement field.

Calculate the state of strain at the origin; at the point, x,y.

Respond again but now with $u(x,y) = \frac{\alpha}{2} \cdot y$ and $v(x,y) = \frac{\alpha}{2} \cdot x$

7

Material Properties and Failure Phenomena

Up until now we have not said much about the way structures behave in the so-called "real world". Our focus has been on abstract concepts and pictures, with generous use made of mathematics in the formulation and solution of problems. There have been exceptions: We have talked about the properties of a rod, how we could measure the stress at which it would yield. We have talked about linear force/deflection relations and how we could measure the rod's *stiffness, k*. But our elaboration and application of the principles of equilibrium and compatibility of deformation have not required any reference to the things of the material world: Compatibility of deformation is a matter of the displacement of points, their absolute and relative displacement (when we talk about strain). Equilibrium of force and of moment concerns concepts that are just as abstract as displacements of points and rotations of infinitessimal line segments - if not more so. Constructing a free body diagram is an abstract, intellectual activity. No one goes out and actually cuts through the truss to determine the forces within its members.

In this chapter, we confront the world of different structural materials and their actual behavior. We want to know how the stiffness, *k*, depends upon the actual material constitution of our beam, or truss member, or concrete mix. We want to know how great a weight we can distribute over the beam or hang from the nodes of a truss before failure. We summarize our interests with two bullets:

- What properties characterize the behavior of a linear, elastic structural element? What is the general form of the stress/strain relations for an isotropic continuum?
- What conditions can lead to failure of a structure?

We begin with our elaboration of the constitutive relations for a continuum.

7.1 Stress/Strain Relations

We want to develop a set of stress/strain relations for a continuous body, equations which apply at each and every point throughout the continuum. In this we will restrict our attention, at least in this chapter, to certain types of materials, namely *homogeneous, linear, elastic, isotropic bodies*.

- *Homogeneous* means that the properties of the body do not vary from one point in the body to another.

- *Linear* means that the equations relating stress and strain are linear; changes in stress are directly proportional to changes in strain (and the other way around, too).

- *Elastic* means that the body returns to its original, undeformed configuration when the applied forces and/or moments are removed.

- *Isotropic* means that the stress strain relations do not change with direction at a point. This means that a laminated material, a material with a preferred orientation of "grains" at the microscopic level, are outside our field of view, at least for the moment.

We have talked about *stress at a point*. We drew a figure like the one at the right to help us visualize the nature of the normal and shear components of stress at a point. We say the *state of stress is fully specified by* the normal components, σ_x, σ_y σ_z and the shear components $\sigma_{xy}=\sigma_{yx}$, $\sigma_{yz}=\sigma_{zy}$ $\sigma_{xz}=\sigma_{zx}$.

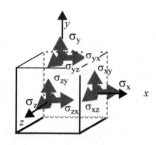

With these restrictions and a heavy dose of symmetry, we will be able to construct a set of stress/strain equations that will apply to many structural materials. This we do now, performing a sequence of thought experiments in which we apply to an element of stuff at a point each stress component in turn and imagine what strains will be engendered, which ones will not. Again, symmetry will be crucial to our constructions. We start by applying the normal stress component σ_x alone.

We expect to see some extensional strain ε_x. This we take as proportional to the normal stress we apply, in accord with the second bullet above; that is we set

$$\varepsilon_x = \sigma_x/E$$

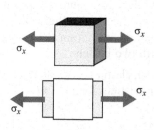

In this we have made use of another bit of real experimental evidence in designating the constant of proportionality in the relationship between the extensional strain in the direction of the applied normal stress to be the *elastic, or Young's modulus, E.* We might not anticipate normal strains in the other two coordinate directions but there is nothing to rule them out, so we posit an ε_y and an ε_z.

Now ε_y and ε_z, because of the indifference of the material to the orientation of the *y* and the *z* axis,– that is, from symmetry– must be equal. We can say nothing more on the basis of our symmetrical thoughts alone.

At this point we introduce another real piece of experimental data, namely that the material *contracts* in the *y* and *z* directions as it extends in the *x* direction due to the applied σ_x. We write then, for the strains due to a σ_x:

$$\varepsilon_y = \varepsilon_z = -\upsilon\varepsilon_x = -\upsilon \cdot \sigma_x/E$$

The ratio of the lateral contraction in the *y* and *z* directions to the extension in the x direction, the so called *Poisson's ratio* is designated by the symbol v. We have encountered the magnitude of the elastic modulus E for 1020, cold rolled steel in the previous chapter. Poisson's ratio, v is new; it takes on values on the order of one-quarter to one-half, the latter value characterizing an[1] incompressible material.

But what about the shear strains? Does σ_x engender any shear strains? The answer is no and here symmetry is all that we need to reach this conclusion. The sketch below shows two possible configurations for the shear strain γ_{xy}. Both are equally possible to an unbiased observer. But which one will follow the application of σ_x?

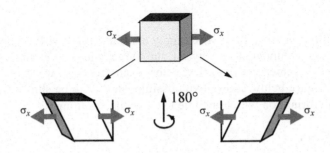

There is no reason why one or the other should occur.[2] Indeed they are in contradiction to one another; that is, if you say the one at the left occurs, I, by running around to the other side of the page, or more easily, by imagining the bit on the left rotated 180° about a vertical axis, can obtain the configuration at the right. But this is impossible. These two dramatically different configurations cannot exist at the same time. Hence, neither of them is a possibility; a normal stress σ_x will not induce a γ_{xy}, or for that matter, a γ_{xz} shear strain.

By similar symmetry arguments, not provided here, we can rule out the possibility of a γ_{yz}. We conclude, then, that under the action of the stress component σ_x alone, we obtain only the extensional strains written out above.

Our next step is to apply a stress component σ_y alone. Now since the body is isotropic, it does not differentiate between the x and y directions. Hence our task is easy; we simply replace x by y (and y by x) in the above relationships and we have that, under the action of the stress component σ_y alone, we obtain the extensional strains

$$\varepsilon_y = \sigma_y/E \qquad\qquad \varepsilon_x = \varepsilon_z = -\upsilon(\sigma_y/E)$$

---

1. In fact, Poisson proved that, for an isotropic body, Poisson's ratio should be exactly one-quarter. We claim today that he was working with a faulty model of the continuum. For some relevant history on early nineteenth century developments in the continuum theories see Bucciarelli and Dworsky, *SOPHIE GERMAIN, an Essay in the Development of the Theory of Elasticity.*
2. Think of the icon at the top as Buridan's ass, the two below as bales of hay.

The same argument applies when we apply the normal stress σ_z alone.

Now if we apply all three components of normal stress together, we will generate the extensional strains, and only the extensional strain.

$$\varepsilon_x = (1/E) \cdot [\sigma_x - \nu(\sigma_y + \sigma_z)]$$
$$\varepsilon_y = (1/E) \cdot [\sigma_y - \nu(\sigma_x + \sigma_z)]$$

and

$$\varepsilon_z = (1/E) \cdot [\sigma_z - \nu(\sigma_x + \sigma_y)]$$

One possibility remains: What if we apply a shear **stress**? Will this produce an extensional strain component in any of the three coordinate directions? The answer is no, and symmetry again rules. For example, say we apply a shear stress, σ_{xy}. The figure below shows two possible, shortly to be shown impossible, geometries of deformation which include extensional straining.

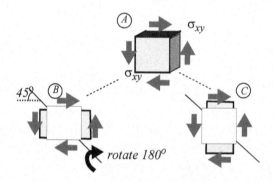

Now I imagine rotating the one on the left about an axis inclined at 45° as indicated. I produce the configuration on the right. Try this with a piece of rectangular paper, a 3 by 5 card, or the like. But this is an impossible situation. The two configurations are mutually contradictory. A like cause, in this case a positive shear stress at the point, should produce a like effect. This is not the case. Hence, neither the deformation of B nor of C is possible.

There remains one further possibility: that a σ_{xy} generates an extensional strain in the x direction equal to that in the y direction. But this too can be ruled out by symmetry[3]. We conclude then that the shear strain σ_{xy}, or σ_{yz} or σ_{xz} for that matter, produces no extensional strains.

The expressions for the extensional strains above are not quite complete. We take the opportunity at this point to introduce another quite distinct cause of the

---

3. This is left as an exercise for the reader.

deformation of solids, namely a *temperature change*. The effect of a temperature change- say ΔT - is to produce an extensional strain proportional to the change. That is, for an isotropic body,

$$\varepsilon_{x \text{ or } y \text{ or } z} = \alpha \Delta T$$

The *coefficient of thermal expansion*, α, has units of $1/^{\circ}C$ or $1/^{\circ}F$ and for most structural materials is a positive quantity on the order of 10^{-6}. Materials with a *negative coefficient of expansion* deserve to be labeled *exotic*. They are few and far between.

The equations for the extensional components of strain in terms of stress and temperature change then can be written

$$\varepsilon_x = (1/E) \cdot [\sigma_x - \nu(\sigma_y + \sigma_z)] + \alpha \Delta T$$
$$\varepsilon_y = (1/E) \cdot [\sigma_y - \nu(\sigma_x + \sigma_z)] + \alpha \Delta T$$
$$\text{and}$$
$$\varepsilon_z = (1/E) \cdot [\sigma_z - \nu(\sigma_x + \sigma_y)] + \alpha \Delta T$$

In the above, we ruled out the possibility of a shear stress producing an extensional strain. A shear stress produces, as you might expect, a shear strain. We state without demonstration that a shear stress produces only the corresponding shear strain. Furthermore, a temperature change induces no shear strain at a point. The remaining three equations relating the components of stress at a point in a linear, elastic, isotropic body are then.

$$\gamma_{xy} = \sigma_{xy}/G$$
$$\gamma_{xz} = \sigma_{xz}/G$$
$$\text{and}$$
$$\gamma_{yz} = \sigma_{yz}/G$$

Recall that $\sigma_{xy} = \sigma_{yx}$. In these, G, *the shear modulus* is apparently a third elastic constant but we shall show in time that G can be expressed in terms of the elastic modulus and Poisson's ratio according to:

$$G = \frac{E}{2(1 + \upsilon)}$$

To proceed, we consider a special instance of a case of plane stress, i.e., one in which the "z components" of stress at a point are zero. The special instance is shown in the figure, at the left.

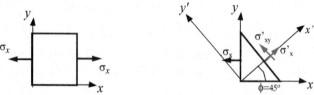

We then consider the stress components acting upon a plane inclined at 45°. We relate the shear stress, σ'_{xy}, on the inclined plane to the normal stress σ_x through the appropriate transformation relation, namely:

$$\sigma'_{xy} = -\left[\frac{(\sigma_x - \sigma_y)}{2}\right] \cdot sin2\phi + \sigma_{xy} cos2\phi$$

With $\phi = 45°$ this gives $\sigma'_{xy} = -\dfrac{\sigma_x}{2}$

Doing the same for the strain component, γ_{xy}',

$$\frac{\gamma'_{xy}}{2} = -\left[\frac{(\varepsilon_x - \varepsilon_y)}{2}\right]sin2\phi + \frac{\gamma_{xy}}{2}cos2\phi$$

gives $\dfrac{\gamma'_{xy}}{2} = -\left[\dfrac{(\varepsilon_x - \varepsilon_y)}{2}\right]$ Now we apply the stress strain relations

$$\varepsilon_x = (1/E) \cdot \sigma_x$$
$$\varepsilon_y = (1/E) \cdot [-v\sigma_x]$$
$$\text{and}$$
$$\gamma'_{xy} = \sigma'_{xy}/G$$

to this last relationship and obtain

$$\frac{\sigma'_{xy}}{2G} = -\frac{(1+v)}{2E}\sigma_x$$

But from the transformation relationship between the stress components above, we know that $\sigma'_{xy} = -\dfrac{\sigma_x}{2}$. For these last two relationships to be consistent, we must have

$$\boxed{G = \frac{E}{2(1+v)}}$$

7.2 Properties of Ordinary Structural Materials

Contrary to my introductory remarks at the outset of this chapter, it seems we have proceeded abstractly in our exploration of the constitution of structural materials. The reason for this is that thinking things through is relatively cheap and inexpensive work compared to doing actual experiments in the real world. If we can figure out some ways in which our materials might, or must, behave by thinking abstractly about continuity and symmetry, about stress and strain, about rotation of axes - all the while making sure our analysis is logical and coherent - we have established a solid basis for fixing the behavior of real materials in the real world. This as long as our materials fit the assumptions of our model as set out in the bullets at the outset of this chapter[4]. Still, it doesn't give us the full picture, the full story; eventually we have to go into the lab to pull apart the actual stuff. To get our hands dirty, we explore how a bar in tension behaves.

Force/Deformation - Uniaxial Tension.

We have already said a few words about the failure of a truss member in tension – how a material like aluminum or steel will begin to *yield* or a more brittle material *fracture* when the tensile *stress* in the member becomes too large in magnitude. We want to say more now; in particular, we want to attend to the deformations that occur in a bar under uniaxial tension and look more closely at the mechanisms responsible for either *brittle fracture or the onset of yield*.

The tension test[5] is a standard test for characterizing the behavior of bars under uniaxial load. The test consists of pulling on a circular shaft, nominally a centimeter in diameter, and measuring the applied force and the *relative displacement* of two points on the surface of the shaft in-line with its axis. As the load P increases from zero on up until the specimen breaks, the relative distance between the two points increases from L_0 to some final length just before separation. The graph at the right indicates the trace of data points one might obtain for load P versus ΔL where

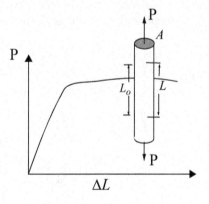

$$\Delta L = L - L_o$$

4. Recall how difficulties arise if there is a misfit - if our model is not appropriate as was the case concerning the behavior of the student in a chair on top of a table tilted up.
5. Standard tests for material properties, for failure stress levels, and the like are well documented in the American Society for Testing Materials, ASTM, publications. Go there for the description of how to conduct a tensile test.

Now, If we were to double the cross-sectional area, A, we would expect to have to double the load to obtain the same change in length of the two points on the surface. That indeed is the case, as Galileo was aware. Thus, we can extend our results obtained from a single test on a specimen of cross sectional area A and length L_0 to another specimen of the same length but different area if we plot the ratio of load to area, the *tensile stress*, in place of P.

Similarly, if, instead of plotting the change in length, ΔL, of the two points, we plot the stress against the *ratio of the change in length to the original length* between the two points our results will be applicable to specimens of varying length. The ratio of change in length to original length is just the *extensional strain*.

We, as customary, designate the *tensile stress* by σ and the *extensional strain* by ε; We assume that the load P is uniformly distributed over the cross sectional area A and that the relative displacement is "uniformly distributed" over the length L_0. Both stress and strain are rigorously defined as the limits of these ratios as either the area or the original length between the two points approaches zero. Alternatively, we could speak of an *average stress* over the section as defined by

$$\sigma \equiv P/A$$

and an *average strain* as defined by

$$\varepsilon \equiv (\Delta L)/L_o$$

The figure below left shows the results of a test of *1020, Cold Rolled Steel*. Stress, σ is plotted versus strain ε. The figure below right shows an abstract representation of the stress-strain behavior as *elastic, perfectly plastic* material.

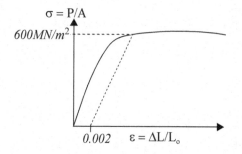 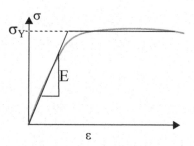

Observe:

- The plot shows a region where the stress is proportional to the strain. The *linear* relation which holds within this region is usually written

$$\sigma = E \cdot \varepsilon$$

where E is the coefficient of *elasticity* of 1020CR steel - 30 x 10^6 lb/in^2.

- The behavior of the bar in this region is called *elastic*. **Elastic means that when the load is removed, the bar returns to its original, undeformed configuration.** That is L returns to L_0. There is no *permanent set*[6].

- The relative displacements of points — the strains in the elastic region — are very small, generally insensible without instruments to amplify their magnitude. To "see" a relative displacement of two points originally 100 *mm* apart when the stress is on the order of 400 *Mega Newtons/m^2* your eyes would have to be capable of resolving a relative displacement of the two points of 0.2 *mm*! Strains in most structural materials are on the order of *tenths of a percent at most*.

- At some stress level, the bar **does not** return to its undeformed shape after removing the load. This stress level is called the *yield strength*. The yield strength defines the limit of elastic behavior; beyond the yield point the material behaves *plastically*. In most materials definition of a nominal value for the yield strength is a matter of convention. Whether or not the material has returned to its original shape upon removal of the load depends upon the resolution of the instrument used to measure relative displacement. The convention of using an offset relies upon the gross behavior of the material but this is generally all we need in engineering practice. In the graphs above, we show the *yield strength* defined at a *2% offset*, that is, as the intersection of the experimentally obtained stress-strain curve with a straight line of slope E intersecting the strain axis at a strain of 0.002. Its value is approximately 600 *MN/m^2*.

- Loading of the bar beyond the yield strength engenders very large relative displacements for relatively small further increments in the stress, σ. Note that the stress is defined as the ratio of the load to the *original area*; once we enter the region of plastic deformation, of *plastic flow*, the bar will *begin to neck down* and the cross sectional area at some point along the length will diminish. The *true stress* at this section will be greater than σ plotted here.

- For some purposes, it is useful to *idealize* the behavior of the material in tension as *elastic, perfectly plastic*; that is, the yield strength fixes the maximum load the material can support. This fantasy would have the material stretch out to infinite lengths once the yield strength was reached. For most engineering work, a knowledge of yield strength is all we need. We design to make sure that our structures never leave the elastic region.[7]

---

6. Note that *linear behavior* and *elastic behavior* are independent traits; one does not necessarily imply the other. A rubber band is an example of an elastic, non-linear material and you can design macro structures that are non-linear and elastic. Linear, inelastic materials are a bit rarer to find or construct.

7. On the other hand, if you are designing energy absorbing barriers, machine presses for *cold rolling or forming* materials and the like, plastic behavior will become important to you.

Exercise 7.1

A steel bolt, of 1/2 inch diameter, is surrounded by an aluminum cylindrical sleeve of 3/4" diameter and wall thickness, t= 0.10 in. The bolt has 32 threads/inch and when the material is at a temperature of $40°C$ the nut is tightened one-quarter turn. Show that the uniaxial stresses acting in the bolt and in the sleeve at this temperature are $\sigma_{bolt} = 79 \ MN/m^2$, and $\sigma_{sleeve} = -63 \ MN/m^2$ where the negative sign indicates the aluminum sleeve is in compression. What if the bolt and nut are cooled; at what temperature might the bolt become loose in the sleeve?

Compatibility of Deformation

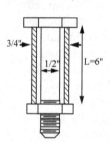

Compatibility of Deformation is best assured by playing out a thought experiment about how the bolt and sleeve go from their initial unstressed, undeformed state to the final state. Think of the bolt and nut being separate from the sleeve. Think then of turning down the nut one quarter turn.

We show this state at the left, below. Δ is the distance traveled in one quarter turn which, at 1/32 *inch/turn* is
$$\Delta = 1/128 \ in$$

Next think of stretching the bolt out until we can once again fit the nut-bolt over the aluminum sleeve, the latter still in its undeformed state. This is shown in the middle figure below. Now, while stretching out the bolt in this way, replace the

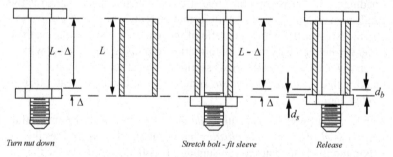

| *Turn nut down* | *Stretch bolt - fit sleeve* | *Release* |

aluminum sleeve[8] then let go. The bolt will strive to return to its undeformed length — the behavior is assumed to be elastic — while the aluminum sleeve will resist contraction. The final state is shown at the right. The net result is that the steel bolt has extended **from its undeformed state** a distance d_b while the aluminum sleeve has contracted a distance d_s. We see from the geometry of these three

---

8. Since this is a *thought experiment* we don't have to worry about the details of this physically impossible move.

figures that we must have, for compatibility of deformation, $d_s + d_b = \Delta$ which is one equation in two unknowns.

Equilibrium

The figure at the right shows an isolation made by cutting through the bolt and the sleeve at some arbitrary section along the axis.

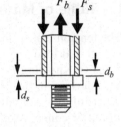

Note in this I have violated my usual convention. I have taken the force in the aluminum as positive in compression.
We let F_s be the **resultant** compressive force in the sleeve, the sum of the distributed loading around the circumference. F_b is the tensile force in the bolt. Like the carton-tie-down exercise, these two internal forces are *self equilibrating*; there are no external applied forces in the final state. We have $\qquad F_b - F_s = 0$

The normal stresses in the sleeve and the bolt are found assuming the resultant forces of tension and compression are uniformly distributed over their respective areas. Equilibrium then can be expressed as $\quad \sigma_b \cdot A_b = \sigma_s \cdot A_s \quad$ where the A's are the cross sectional areas of the bolt and of the sleeve.

Constitutive Relations

The constitutive relations are, for uniaxial loading, which is the case we have on hand,

$$\sigma_s = E_s \cdot (d_s/L) \qquad \text{and} \qquad \sigma_b = E_b \cdot (d_b/L)$$

We have then a total of four equations for four unknowns – the two displacements, the two stresses. Substituting the expressions for the stresses in terms of displacements into the equilibrium allows me to write

$$d_b = d_s \cdot (A_s E_s / A_b E_b)$$

which tells me the relative deformation as a function of the relative stiffness of the two material. If the sleeve is "softer", the bolt deforms less... etc.

With this, compatibility gives me a way to solve for the displacements in terms of Δ. I obtain, letting

$$\beta = (A_s E_s / A_b E_b)$$

we have

$$d_s = \Delta \frac{1}{(1+\beta)} \qquad \text{and} \qquad d_b = \Delta \frac{\beta}{(1+\beta)}$$

Values for the stresses are found to be $\sigma_b = 79 \ MN/m^2$ and $\sigma_s = 63 \ MN/m^2$. (Note: compressive)

In computing these values, the elastic modulus for steel and aluminum were taken as 200 GN/m^2 and 70 GN/m^2 respectively. Observe that, though the steel experiences less strain, its stress level is greater in magnitude than that seen in the aluminum.

7.3 Table of Material Properties

The tension test is the standard test to determine E, the elastic or Young's modulus. Test that load a cylindrical specimen in torsion are used to measure the shear modulus G. Knowing E and G, Poisson's ratio may be obtained from the relationship we derived in the previous section.

What follows is a table giving the elastic properties and failure stresses in tension (and/or compression) for common structural materials. "Failure" means that ordinarily you want to design your structure such that you do not come close to this value under anticipated loading conditions.

The variety of materials included is meant to give the reader some idea of the range of property values of different kinds of structural materials. The values themselves are only meant to indicate orders of magnitude. In some cases, where the range of property values for a material is so large due to differences in composition or quality of its fabrication, a range has been shown. And certainly the table is not meant to be complete, nor should the values be used in detailed design work.

Material	Specific Gravity	Elastic Modulus		Failure Stress		α
		10^6 psi	10^9 N/m^2	10^3 psi	10^6 N/m^2	10^{-6} /°C
Al 2024-T3	2.7	10	70	60	400	23
Al 6061-T6	2.6	10	70	40	280	23
Al 7075-T6	2.7	10	70	80	550	23
Concrete	2.3	3	20	3 - 6[a]	20 - 40	7-12
High Strength	2.3	3 - 5	20 - 35	5-12	35 - 80	7-12
Copper	9	15	100	5	35	16
Glass Fiber	2.7	10	65	2000	15000	8
Iron (cast)	7	15	100	20-40	150-300	10
Steel High Strength	8	30	200	50-150	300-1000	14
Steel Structural	8	30	200	40-100	250-700	12
e.g., AISI C1020	8	30	200	85	600	12
Titanium	5	15	100	100	700	9
Wood (pine)	0.5	1.4	10	1	7	

a. In Compression

In all of this, failure stress is no unique number; in contrast to the elastic modulus, *E*, which is safe to take as a single value across different varieties and compositions of a material[9], the failure stress will vary all over the lot depending upon composition, care and means of fabrication. Compare the yield stress of cold rolled versus hot-rolled 1020 steel. Note too that one does not design to the failure stress but to a level significantly less than the numbers in the table. A *factor of safety* is always introduced to ensure that internal loads in the structure stay well below the failure level.

7.4 Failure Phenomena

Failure comes in different guises, in different sizes, colors, shapes, and with different labels. We have talked about *yielding, the onset of plastic flow* of *ductile* materials - materials which show relatively large, even sensible, deformations for relatively small increases in load once the material is loaded beyond its *yield strength*. If the excessive load is removed before complete collapse, the structure will not wholly return to its original, undeformed configuration.

Although it is the tension test that is used to fix the limit of elastic behavior and to define a *yield strength*, the mechanism for yielding is a shearing action of the material on a microscopic level. We have seen how a tensile stress in a bar can produce a shear stress on a plane inclined to the axis of the bar.

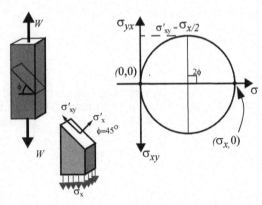

The figure shows the Mohr's Circle for a bar subjected to uniaxial tension. The maximum shear stress occurs on planes oriented at 45° to the axis of the bar. Its value is one-half the applied tensile stress. This then can be taken as a criterion for the onset of yield of a ductile material: Whenever the maximum shear stress at any point within any element of a structure exceeds one-half the applied tensile stress at the yield point in a tension test, the element will yield[10]

Not all materials behave as steel or aluminum or ductile plastics. Some materials are *brittle*. Load glass, cast iron, a brittle plastic, a carbon fiber, or concrete in a tension test and they will break with very little extension. They show insensible

---

9. Concrete is an exception. Indeed it is difficult to find a linear portion of the stress/strain data taken from a specimen in compression.

10. In a general state of stress, the maximum shear stress is given by one-half the maximum difference of the (three) principle stresses. In the uniaxial tension test, this is just one-half the tensile stress as shown in the figure.

deformation all the way up to the fracture load. Recall the exercise in chapter 4 where we subjected a piece of chalk to torsion and how this generated a maximum tensile stress on a plane inclined at 45° to a generator on the surface of the chalk.

Metals generally carry a tensile or compressive load equally well. Concrete can not. Concrete generally can carry but one-tenth its allowable compressive load when subject to tension. A different sort of test is required to measure the tensile strength of concrete. (See the insert).

In an *indirect* tension test of concrete, a cylinder is loaded with a distributed load along diametrically opposed, sides of the cylinder. These line loads engender a uniform tensile stress distributed within the cylinder over the plane section A-A', bisecting the cylinder - except within the vicinity of the circumference. This tensile stress can be shown to be

$$\sigma_x = 2P/(\pi LD)$$

where L is the length (into the page) of the cylinder along which the load P is distributed, and D is the cylinder diameter.

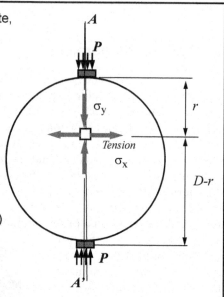

A compressive stress on planes orthogonal to A-A' is also engendered at each point. This can be shown to be equal to

$$\sigma_x = 2P/(\pi LD)\left[\frac{D^2}{r\,(D-r)} - 1\right]$$

Values for the compressive and tensile failure stress levels depend upon the quality and uniformity of the composition of the material. Recall how we assumed our material was homogeneous. If this is not the case, then other features of the material must be taken into account. For example, in a composite material or structural element - a member made up of two different materials - careful attention must be paid to the interface of the different materials out of which the member is made. We will consider some examples of composite structural elements in our chapter on the behavior of beams[11].

Another violation of homogeneity can prove disastrous: If the material, supposed continuous and uniform, contains an *imperfection*, e.g., a microscopic crack, unseen by the naked eye even if on the surface, then all bets are off. (Rather all bets are on!) A crack can be the occasion for the magnification of stress levels in its immediate neighborhood.

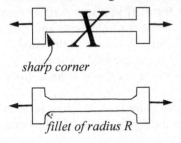

sharp corner

fillet of radius R

This is one good reason why you see *fillets* at corners within a machined part. For example, one would never make a tensile test specimen in the shape shown at the top of the figure. The sharp corners where the two cylinders of differing diameters meet would engender a stress level at that junction significantly greater than the tensile stress found at the middle of the specimen, away from the junction.

A classic example of how a discontinuity within the interior of a material can act as a *stress raiser* is the solution obtained from the theory of elasticity for the stress field around a hole in a thin plate.

The figure at the right shows the effect of a circular hole on the stress component σ_x engendered in a thin plate subjected to a uniform tensile stress in the x direction. We show only the normal stress component σ_x acting on an x face which is a continuation of the vertical diameter of the hole. The hole gives rise to a stress concentration three times the magnitude of the uniformly applied tensile stress[12].

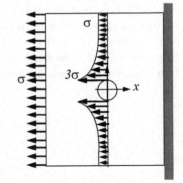

So far in our discussion we have been concerned with static loading conditions, i.e., we determine the internal stresses and static displacements due to loading - such as a dead weight. But some failure phenomena take time to develop. Even if there is no perceptible dynamic displacement or motion, materials age - like you and me - with the passage of time. Material properties and modes of failure also may depend upon temperature. What may be ductile at room temperature will be brittle at low temperatures. At high temperatures, still well below the melting point, materials will *creep* – they will continuously deform at a constant load.

A material can fail in *fatigue*: Under continual cycling through tension then compression, a material will fail well below the yield strength or fracture stress

11. Think of fiber reinforced skies, poles for pole vaulting, frames of tennis racquets and steel reinforced concrete slabs or beams.
12. Reference: Timoshenko and Goodier, *Theory of Elasticity*, McGraw-Hill, N.Y. Third Edition 1970.

witnessed in a tension test. Here the number of cycles may be large, on the order of thousands. But dynamic excitation of a structure can lead to failure in but a few cycles if the excitation drives a resonant mode of the system. Think of the Tacoma Narrows Bridge where aerodynamic excitation at the resonant frequency of a torsional mode of the structure led to ever increasing amplitude of vibration and the eventual spectacular collapse of the span. Before this failure event, aerodynamics and structures were the concerns of two different worlds[13].

Failure often occurs where you are not looking for it. If you spend all of your time doing models of structures made up of a large number of elements and focus solely on stress levels within the elements and pay little attention to the way the members are joined together and/or fixed to ground you are headed for trouble, for failure often occurs at joints. The Hyatt Regency walkway collapse is an unfortunate example.

Some failure phenomena are "macro"; they require more than the consideration of the state of stress at a point. *Buckling* is of this nature. We will study the buckling of beam-columns in the last chapter.

When we address the possibility of failure, inspection and testing become a necessity. Here we move from the abstract world of the theory of elasticity to the world of empirical data, manufacturers' specifications, of codes and traditional ways of fabrication and assembly. The problem is that in the design of the new and innovative structure, there is always the possibility that the codes and regulations and traditional ways of ensuring structural integrity do not exactly apply; something differs from the norm. If after a careful reading of existing code, any question remains, a full test program might be called for.

---

13. Except within the world of aerospace engineering where the coupling of aerodynamics and structural behavior were well attended to in the design of airfoils.

Design Exercise 7.1

A solid circular steel shaft of diameter 40 *mm* is to be fitted with a thin-walled circular cylindrical sleeve, *also made of steel*. In service the system is to serve as a stop, halting the motion of another fitted, but freely moving cylindrical tube whose inner radius is slightly larger than the outer radius of the solid shaft. The stopping sleeve is to remain in fixed location on the solid shaft for all axial loads less than some critical value of the force *F* shown in the figure. That is, for *F*< *50kN*. If *F* exceeds this limit the sleeve is to frictionally break free and allow the sliding cylinder to continue moving along the shaft.

It is proposed to fasten the sleeve to the shaft by means of a *shrink fit*. The initial inner radius of the sleeve is to be made slightly *less* than the initial outer radius of the shaft. The sleeve is then heated to a temperature not to exceed $\Delta Tmax = 250^{\circ}C$ so that its heat-treatment is not affected. The *hot* sleeve is then slipped over the shaft and positioned as desired. When the sleeve cools down, the radial *misfit* between the shaft outer radius and the sleeve's unstressed inner radius will generate sufficient mechanical interaction between the two so that the stopping and break-away functions can be fulfilled.

Size the sleeve.

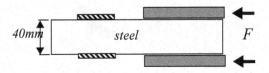

7.5 Problems

7.1   Hoop #1 is enclosed within hoop #2. The two are made of different materials, have different thicknesses but the same width (into the page). They are shown in their unstressed state, just touching. *Show that* after tightening the bolt at the top of the assembly and closing the gap, Δ, to zero, the stress in the outer hoop is tensile and has magnitude $F/(bt_1)$ while the stress in the inner hoop is compressive and has magnitude $F/(bt_2)$. In these t_1 and t_2 are the thicknesses,

$$F = k_1 k_2 \Delta/(k_1 + k_2)$$

where

$$k_1 = (bt_1)E_1/L_1 \qquad and \qquad k_2 = (bt_2)E_2/L_2$$

What if an internal pressure is applied to the inner hoop? When will the stress in the inner hoop diminish to zero? What will be the hoop stress in the outer hoop at this internal pressure?

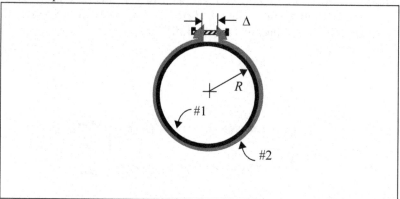

7.2   The thin plate is a composite of two materials. A quarter-inch-thick steel plate is clad on both sides with a thin ($t_{al} = 0.005$ in), uniform, layer of aluminum. The structure is stress-free at room temperature. *Show that* the stresses generated in the two materials, when the temperature changes an amount ΔT, may be approximated by

$$\sigma_{al} = (\alpha_{st} - \alpha_{al})E_{al}\,\Delta T/(1-\nu) \quad and \quad \sigma_{st} = -(2t_{al}/t_{st})(\alpha_{st} - \alpha_{al})E_{al}\Delta T/(1-\nu)$$

At what temperature will the clad plate begin to plastically deform? Where?

7.3 Two cylindrical rods, of two different materials are rigidly restrained at the ends where they meet the side walls. The system is subject to a temperature increase ΔT.

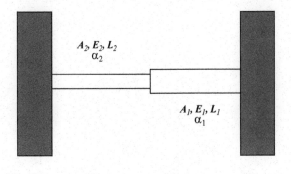

How must their properties be related if the point at which they meet is not to move left or right?

If material #1 is steel and #2 is aluminum, what more specifically can you say?

$E_1 = 200$ GPa   steel

$E_2 = 70$ GPa   aluminum $\alpha_1 = 15$ e-06 /$^{\circ}$C $\alpha_2 = 23$ e-06 /$^{\circ}$C

8

Stresses/Deflections Shafts in Torsion

8.1 An Introductory Exercise

We return to the problem of torsion of circular shafts. We want to develop methods to determine the *shear stress distribution over the cross-section* of the *torque-bearing structural element* and the *rotation of any cross-section* relative to another. Although we limit our attention to circular cross-sections, this ought not to be taken to imply that only circular shafts are available to carry *torsional loads*. For example:

Exercise 8.1

A single bay of a truss structure, typical of the boom of a construction crane, is shown below. Show that the torsional stiffness of the section 1-2-3-4 relative to the other, fixed section, is given by

$$M_T = K_T \cdot \phi \qquad where \qquad K_T = 2(AEa) \cdot (\cos\alpha)^3$$

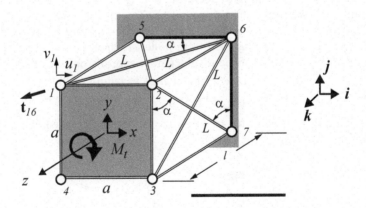

In this, $\cos\alpha = a/\sqrt{(a^2+l^2)}$ and ϕ is the rotation of the section 1-2-3-4.

To determine a *stiffness* we must necessarily consider displacements of the structure at the four, *unconstrained* nodes. But since we are only interested in the *torsional stiffness*, that is, the relation between a torque applied at the free end section and the rotation of that same section, our task is lighter.

Our method of analysis will be a *displacement formulation*. That is, we will impose a *displacement field* of a particular kind, namely a *rigid body rotation* of the plane section 1-2-3-4 about the z axis of the structure and determine what forces are required to be applied at the nodes to maintain the displaced configuration. Rigid here means that none of the truss members lying in the plane of this section experience any change in length; their end nodes retain their same **relative** positions - the square section remains a square.

The figure shows the imposed rotation about the z axis; the rotation ϕ is understood to be *small* in the sense that the magnitude of the displacements of the nodes can be taken as the product of the half-diagonal of the section and the angle. Vector expressions for the displacements are then:

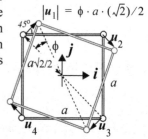

$$\mathbf{u}_1 = (a/2)\phi\mathbf{i} + (a/2)\phi\mathbf{j} \qquad \mathbf{u}_2 = (a/2)\phi\mathbf{i} - (a/2)\phi\mathbf{j}$$

$$\mathbf{u}_3 = -(a/2)\phi\mathbf{i} - (a/2)\phi\mathbf{j} \qquad \mathbf{u}_4 = -(a/2)\phi\mathbf{i} + (a/2)\phi\mathbf{j}$$

Now if the truss members lying in the plane of the section do not deform, they do not offer any resistance to rotation. But then what members **do** resist this particular *displacement field* and provide the torsional stiffness?

We inspect them all. Consider member 1-5: If we do not allow a z displacement, its change in length will be zero. It is zero because, even though node #1 displaces relative to node #5, the displacement vector of #1 is perpendicular to the member; the projection of the displacement upon the member is zero. The same can be said about members 2-6, 3-7, and 4-8.

The same **cannot** be said of member 1-6, or any of the other diagonal members. The displacement of node #1 **does** have a non-zero projection upon the member 1-6, a projection tending to decrease its length when ϕ is positive as shown. To determine the magnitude of the contraction (as well as certifying that indeed the member does contract) we proceed formally, constructing a unit vector along the member in the direction *outward* from node #1, then determine the projection by taking the scalar product of the displacement and this unit vector. The unit vector is $\mathbf{t}_{16} = -cos\alpha \cdot \mathbf{i} + sin\alpha \cdot \mathbf{k}$. The change in length of the member is then

$$\delta_{16} = \mathbf{t}_{16} \bullet \mathbf{u}_1 = -(a/2)\phi cos\alpha$$

Less formally, we can try to visualize how the displacement **does** produce a contraction. We show a top view:

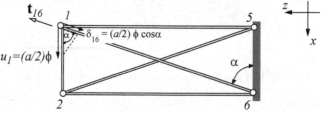

We note that, of the two scalar components of $\boldsymbol{u}_1$, the y component, $(a/2)\phi\,\boldsymbol{j}$ - $\boldsymbol{j}$ out of paper - will be perpendicular to the member and hence it will not change the member's length in any way. The same cannot be said for the other component, the x component $(a/2)\phi$ i.

Hopefully it is clear that the projection of this component along the member is just $(a/2)\phi\cos\alpha$ as obtained from the dot product. Note that it acts to shorten member 1-6. Similarly, the other diagonal members, 2-7, 3-8, and 4-5 will also contract and by the same amount.

Using a similar argument, we can show that members 2-5, 3-6, 4-7 and 1-8 will **extend** the same amount.

Compressive and tensile forces will then be engendered in the diagonals by this particular displacement field and it is these which resist the applied torque.

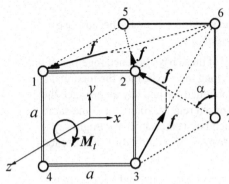

To see how, we make an isolation of the sections and require moment equilibrium about the z axis.

First note that force equilibrium in the three coordinate directions x, y, *and z* is satisfied: Since all eight diagonal members experience the same magnitude of extension or contraction the tensile or compressive force each will experience will be of the same magnitude. So at each node, the z component of one in tension will just balance the z component of the another in compression.

We denote the value of the tensile load by f. Assuming a linear, elastic force-deformation relationship for all diagonal members of cross-sectional area A and elastic modulus E, we can write:

$$f = (AE/L) \cdot \delta$$

The magnitude of the extension or contraction of each diagonal is the same

$$\delta = \delta_{16} = (a/2) \cdot \phi \cdot \cos\alpha$$

so $\quad f = \left(\dfrac{AE}{L}\right) \cdot \delta = \left(\dfrac{AE}{L}\right) \cdot (a/2) \cdot \phi \cdot \cos\alpha = (AE/2) \cdot \phi \cdot \cos\alpha^2$

where we have used the fact that $\quad L = \dfrac{a}{\cos\alpha}$

Now **each** of the diagonal members produces a moment about the z axis. More precisely, the **horizontal** components of the member forces in the top and bottom diagonals and the **vertical** components of the member forces in the diagonals along the sides each contributes a moment of magnitude $f\cos\alpha$ $(a/2)$. All eight together then provide a torque resisting the imposed rotation ϕ of

$$\boxed{M_T = 4af \cdot \cos\alpha = 2aAE(\cos\alpha)^3 \cdot \phi}$$

Here, then, is a structure capable of resisting torsion about its axis. Observe that:

- The units check. AE has the units of force since E, the elastic modulus, has the units of stress or force per unit area: a has the dimensions of length so the right-hand side has the dimensions of the left, that of force times length, that of a moment or torque.

- As we increase the length of the bay relative to the length of the side of the square cross-section, the angle α increases, $\cos\alpha$ decreases, and the torsional stiffness decreases - and dramatically since *it goes as the cube of the cosine*.

- The relationship between torque and rotation ϕ is *linear* because we assumed *small displacements and rotations*.

- Imagine a structure built up of many of these bays fastened together. If this structure includes ten bays aligned along the z axis, then the torsional stiffness would be one-tenth the value obtained for one bay alone since the over-all rotation would be the sum of ten relative rotations of the same magnitude. The torque acting on each bay is the same. We might say the total angular rotation is *uniformly distributed* over the full string of ten bays.

- We can go back and determine the forces and/or normal stresses in the diagonal members given an applied torque M_T. We have:

$$M_T = 4af \cdot \cos\alpha \qquad \text{so} \qquad f = M_T/(4a\cos\alpha)$$

- Finally, the structure is redundant. We could remove some members and still carry the load. But note that if we remove one of diagonal members from each of the four sides of the bay our equilibrium requirement in the z direction would not be satisfied by the force system posed. We would have to apply a force in the z direction at each node in order to maintain our prescribed displacement field. Without this additional constraint, the nodes would displace in the z direction and in an un-rotationally symmetric fashion. Our analysis would not go through.

In analyzing the torsion of a circular shaft we will proceed much the same way as above. We will first consider deformations due to a relative rotation of two sections of the shaft and, on the basis of symmetry, construct a compatible strain state. The stress-strain equations give a corresponding stress distribution— one which consists solely of a shear stress acting in the plane of the cross-section. Equilibrium, repeating the maneuvers of a previous chapter, then brings the applied torque into the picture and we end with an equation relating the applied torque to the rotation of the shaft, a *stiffness relation* in form like the one derived above for the truss bay.

8.2   Compatibility of Deformation

The cross-sections of a circular shaft in torsion rotate as if they were rigid in-plane. That is, there is no relative displacement of any two, arbitrarily chosen points of a cross section when the shaft is subjected to a torque about its longitudinal, z, axis. We prove this assertion relying on *rotational symmetry* and upon the constancy of the internal torque as we move down the axis of the shaft.

We first show that radial lines must remain straight by posing that they deform, then show that a contradiction results if we do so.

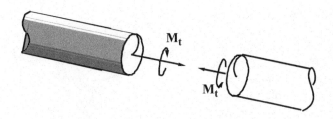

Let the points along a radius take the shape of a curve in plane in the deformed state: Now consider the same set of points but from the perspective of someone who has the portion and the shaft to the right to observe. He or she would necessarily see the deformed locus as shown if the same moment is to have the same effect. Now, however, the two cannot be put back together without leaving a hole within the interior. The two displacement fields are incompatible.

The only displacement pattern that **will** fit back together is if the locus is again a straight line.

It is still possible that, while radial lines remain radial, there could be some sort of accordion effect as we march around the axis of the shaft – some radial lines coming closer together, others widening the angle between them. But no, this is not possible since we have complete rotational symmetry. Whatever happens at one angular position must happen at every other angular position.

There remains the possibility of *out of plane* bulging-out and/or dishing-in. While these would not violate rotational symmetry, we rule them out using the following argument based upon the fact that the torque does not vary as we move along the axis of the shaft.

We posit a bulging-out on a section of shaft when we apply a clock-wise moment. Running around to the other end of the section, we would claim a bulging-out there too, since the moment is again directed along the axis of the section in the same sense. But now consider the portion of the shaft to either side of the cut section. To be consistent, their cross-sections will bulge out. This is clearly an incompatible state of deformation. On the other hand, if it dished-in, then we would have a torque, the same torque producing two dramatically different effects. This argument also rules out a uniform extension or contraction of the cross-section.

Thus, no bulging-out, no dishing-in, radial lines remain radial - a cross-section rotates as a rigid plane.

One further fact follows from the uniformity of torque at each section, namely, **the relative rotation of two cross-sections is the same for any two sections separated by the same distance along the axis of the shaft.**

If we let ϕ be the rotation of any section, then this is equivalent to saying

$$d\phi/dz \text{ is a constant}$$

Consider now the strains due to the rotation of one section relative to another.

The figure shows the rotation of a section located along the axis at $z+\Delta z$ relative to a section at z just below it. Of course the section at z has rotated too, most likely. But it is the relative rotation of the two sections which gives rise to a strain, a shear strain. γ, which measures the decrease in right angle, originally formed by two line segments, one circumferential, the other axially directed as shown. From the geometry we can state:

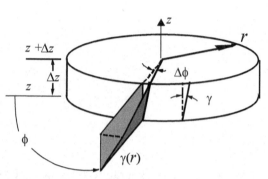

$$\gamma = \lim_{\Delta z \to 0} (r\Delta\phi / \Delta z) = r\frac{d\phi}{dz}$$

Note that this relationship shows that the shear strain is a linear function of radius - zero at the axis, maximum at the shaft's outer radius. Note too that, with $d\phi/dz$ a constant, the shear strain does not vary with z with position along the axis of the shaft.

There are no other strains! With no deformation in plane and no bulging-out or dishing in, there are no other strains. If there existed some asymmetry like that of the truss bay structure with one diagonal number removed from each bay, then we would not be able to rule out a contraction (or extension) in the z direction.

8.3   Constitutive Relations

Because we have but one strain component, this will be a very short section. The corresponding stress is the shear stress τ and is related to the shear strain according to:

$$\tau = G\gamma = G \cdot r\left(\frac{d\phi}{dz}\right)$$

8.4 The Torque-Rotation Stiffness Equation

Back in Chapter 3, we obtained an expression for resultant torque about the axis of a circular shaft due to a shear stress distribution $\tau(r)$, an arbitrary function of r.

We obtained
$$M_T = 2\pi \int_0^R \tau(r)r^2 dr$$

Now, with our linear function of r, we can carry out the integration. Doing so we obtain a relationship between the applied torque M_T and the **rate of rotation**

$d\phi/dz$, a *stiffness* relation.
$$M_T = 2\pi \int_0^R G\frac{d\phi}{dz}r^3 dr$$
or, since $d\phi/dz$ and G are constants,

we are left with the integral of r^3 and can write

$$M_T = GJ\left(\frac{d\phi}{dz}\right)$$

where J is a function only of the geometry of the cross-section - its radius R. You may have encountered it as the *polar moment of inertia* $J = \int_{Area} r^2 dArea$ For the solid circular shaft $J = \pi R^4/2$.

This stiffness equation is analogous to the stiffness relationship derived for one bay of the truss structure considered at the outset of the chapter. For a shaft of length L, the rotation of one end relative to the other is just the integral of the **constant** rate of rotation over the length, that is, just the product of the two. We

obtain then:
$$M_T = (GJ/L) \cdot \phi$$

This is our major result. Observe

- We can obtain the shear stress and strain distribution in terms of the applied moment by substitution. We obtain

$$\tau(r) = r \cdot (M_T/J)$$
$$\gamma(r) = r \cdot M_T/(GJ)$$

- Our analysis is identical for a hollow shaft. All of the symmetry arguments apply. Only the expression for J changes: It becomes

$$J = \int_{Area} r^2 dA = (\pi/2)(R_o^4 - R_i^4)$$

where R_0 is the outer radius and R_i the inner radius of the shaft.

- If we do anything to destroy the rotational symmetry, all bets are off. In particular if we slit a hollow tube lengthwise we dramatically decrease the torsional stiffness of the tube.

If we have a *composite shaft* of two *concentric shafts,* or more, the analysis will go through as follows:

The symmetry arguments still apply; the strain as a function of radius remains linear and proportional to the rate of twist $\gamma = r d\phi/dz$

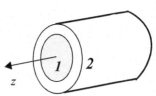

But now our shear stress distribution is no longer the same. Within region 1, the core shaft we have

$$\tau_1 = G_1 \cdot r \cdot \frac{d\phi}{dz} \text{ and } \tau_2 = G_2 \cdot r \cdot \frac{d\phi}{dz} \text{ within region 2, the concentric outer shaft.}$$

The equivalent moment is then $M_T = \int_{A_1} r\tau_1 dA_1 + \int_{A_2} r\tau_2 dA_2$ which yields

$$M_T = [G_1 J_1 + G J_2]\left(\frac{d\phi}{dz}\right) = (\overline{GJ})\left(\frac{d\phi}{dz}\right)$$

The shear strain distribution is then $\gamma = rM_T(\overline{GJ})$ with the shear stress distribution within each region given by

$$\tau_1 = [G_1/(\overline{GJ})]rM_T \qquad \tau_2 = [G_2/(\overline{GJ})]rM_T$$

Note that, although $R_2 > R_1$ the maximum shear stress might occur at the outermost radius of the inner shaft if $G_1 > G_2$.

Exercise 8.2

A torque M_T of 20 Nm is applied to the steel shafts geared together as shown. Now show that

- The internal force acting between the gear teeth is 800. N.
- The maximum shear stress due to torsion is $204 MN/m^2$ and occurs in the smaller diameter shaft at its outer radius.

- The torsional stiffness, K_ϕ, at the end where M_T is applied is 44.35 *Nm/radian..*

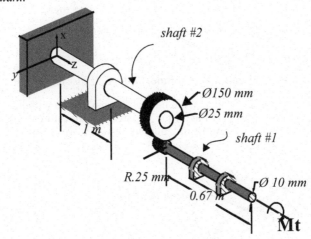

Reid: Hey Katie, you do the first part and I'll do the second, ok?

Katie: Well, I don't know; usually when the author puts out a sequence of questions like this it's best to go through step by step.

Reid: That's only for the do-do's who need their hand held. See, I got the second question wired. It goes like this: the torque is the same in both, right? And the radius of the shaft that's fixed to the wall is biggest and because the maximum shear stress in a shaft is at the biggest radius, I just plug into the formula for shear stress and I've got it.

Katie: I think you're wrong Reid, I think we should go slower, take things step by step.

Reid: You're just jealous because you girls can't see things as quick as us guys.

Katie: It appears that you're so quick you didn't even need to read the problem ..... it says right there that the maximum shear stress occurs in the smaller diameter shaft!

Reid: Maybe the answer given is wrong. Stuff in these books is wrong lots of times. I mean look at that equation for shear stress

$$\tau = rM_T/J.$$

See that little old radius there up top? Now when that *r* maxes out so does τ.

Katie: But Reid, that *little old J* depends on *r* too. In fact, for a solid shaft it goes as the fourth power to the radius so it predominates.

Reid: *Predominates....* You're saying the smaller radius shaft has maximum shear stress for the same moment?

Katie: Hold on, hold on. The maximum shear stress depends upon the moment in the shaft as well.. that "little old" M_T right there. So we have got to first see which shaft carries the largest torque, and then...

Reid: But the torques are the same in both shafts!

Katie: ...same in both shafts. Where do you get that from?

Reid: I dunno...conservation of torque, I guess...it just feels right.

Katie: You men are all alike. What do feelings have to do with it? Watch me: I isolate the shaft carrying the torque M_T, just like the author did with that interesting historical example of a human powered, well-water lift. See, I draw this figure:

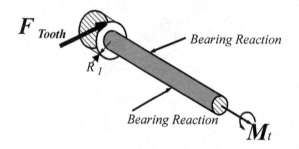

Now moment equilibrium about the axis of the shaft gives me that force on the tooth is going to be

$F_{Tooth} = M_T/R_1$ where R_1 is the radius of the small gear.

Reid: So.....

Katie: Well, **then** I turn to the other shaft and find that the torque is $M_2 = R_2 F_{Tooth} = (R_2/R_1)M_T$, which, my friend, is not equal to, but **three times** the applied torque, M_T. So, Reid, you see how your feelings can lead you astray?

****

Katie is correct, as is the book. The torque in the shaft fixed to the wall is three times the torque carried by the free-ended shaft. Even so, because the shear stress is more sensitive to changes in the radius of a shaft than to torque, it is the smaller shaft that sees the largest shear stress. If both the shafts had the same radius, the one fixed to the wall would indeed show the largest shear, but that is not the case here. I could make up another story about how, with less of a radius, the smaller cross-section must experience a greater stress at each and every position r in order to sum to the same moment.

The work follows:

The maximum stress in shaft #2 is $\tau_2|_{max} = r_2 M_2 / J_2$ which, with $J = \pi \, r_2^4/2$ gives

$$\tau_2|_{max} = 39.2 \; MN/m^2$$

while for shaft #1 we obtain $\tau_1|_{max} = 204 \; MN/m^2$

The torsional stiffness at the end where the movement is applied is obtained by summing up the relative angular deflections of the two shafts being careful to assure compatibility of deformation at the section where the gears mesh. Working from the wall out, the angular rotation of the geared end of shaft #2 relative to its (fixed) end is $\quad \phi_2(L_2) = (M_2 L_2)/(GJ_2)$

Now, compatibility of deformation, no-slip at the gear teeth requires that

$$\phi_2(L_2)R_2 = \phi_1(0)R_1$$

where the directions of the angular rotations are indicated in the figure. So we have

$$\phi_1(0) = (R_2/R_1)(M_2 L_2)/(GJ_2)$$

Now the **relative rotation** of the **free end** of the small shaft to which the torque is applied is given by

$$\phi_1(L_1) \; - \; \phi_1(0) = (M_1 L_1)/(GJ_1)$$

So

$$\phi_1(L_1) = (M_1 L_1)/(GJ_1) + (R_2/R_1)(M_2 L_2)/(GJ_2)$$

Substituting for M_1, M_2 we can write

$$\phi_1(L_1) = (M_1 L_1)/(GJ_1)[1 + (R_2/R_1)^2(L_2/L_1)(J_1/J_2)]$$

So $K\phi = (GJ_1/L_1)/[1 + (R_2/R_1)^2(L_2/L_1)(J_1/J_2)]$ which gives $K\phi = 44.35 \; NM/radian$

Observe:

- Although only numerical results were required to be verified, I worked through the problem symbolically refraining from *plugging-in* until the end. This is highly recommended practice for it allows me to keep a check on my work by inspecting the dimensions of the results I obtain.

- Note, too, how I can readily extend my results to other configurations - changing the relative length of the shafts, their radii, and the material out of which they are made.

- The maximum shear stress is quite large relative to the yield stress of steel. We shall see that yielding becomes a possibility when the maximum shear stress is one-half the yield strength in a tension test.

8.5   Torsion of other than circular shafts

We have already stated how, if a shaft does not exhibit symmetry, our analysis of deformation does not apply. A slit hollow tube, for example, behaves dramatically different from a continuous hollow tube.

Considerable effort has been expended by applied mathematicians over a century or two in developing solutions for the torsion of shafts of other than circular cross section. The results for a shaft sporting a rectangular cross section are well documented[1]. The figure at the right is a classic.

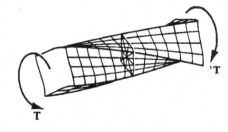

The distorted shape – note the *warping* of the originally plane cross section – is due, in part, to the fact that the shear stresses at the corners must vanish.

Other results say that the maximum shear stress occurs in the middle of the longest side of length b. Its magnitude and that of the torsional stiffness are neatly summarized below. $\tau|_{max} = M_t /[k_2(2a)^2(2b)]$

is the maximum shear stress. And $M_t = K_t \phi/L$ *where* $K_t = k_1 G(2a)^3(2b)$

is the torsional stiffness. Note: a and b are the dimensions of the cross-section. Values for the constants k_1 *and* k_2 as functions of the ratio of the lengths of the sides of the cross section are given in the table below, again taken from Timoshenko and Goodier.

TABLE OF CONSTANTS FOR TORSION OF A RECTANGULAR BAR

$\dfrac{b}{a}$	k	k_1	k_2	$\dfrac{b}{a}$	k	k_1	k_2
1.0	0.675	0.1406	0.208	3	0.985	0.263	0.267
1.2	0.759	0.166	0.219	4	0.997	0.281	0.282
1.5	0.848	0.196	0.231	5	0.999	0.291	0.291
2.0	0.930	0.229	0.246	10	1.000	0.312	0.312
2.5	0.968	0.249	0.258	∞	1.000	0.333	0.333

Reprinted with permission from S.P. Timoshenko and J.N. Goodier, *Theory of Elasticity,* (1970, McGraw-Hill, Inc.)

---

1. The results presented here are drawn from *the* classic text about stress and strain – Timoshenko and Goodier, *THEORY OF ELASTICITY,* Third Edition, McGraw-Hill, 1970.

Design Exercise 8.1

A hollow aluminum shaft, two meters long, must transmit a torque of 20 *KNm*. The total angle of twist over the full length of the shaft is not to exceed 2.0^o, and of course, we do not want the shaft to yield. Size the outside and inside diameters of the shaft.

8.6   Problems – Torsion of Circular Shafts

8.1   *What if* a solid circular shaft is replaced by a square shaft whose **diagonal** is equal to the diameter of the original circular shaft; How does the torsional stiffness change; For the same torque, how does the maximum shear stress change?

What if the solid circular shaft is replaced by a square shaft whose **side** is equal to the diameter of the original circular shaft; How do these change?

8.2   *Estimate* the maximum shear stress in the square, center-post shaft of the drive transmission system shown below when the horse is delivering α *horsepower* where α < 1.0. (1 *horsepower* = 746 *Watts* = 550 *ft-lb/sec*) Estimate the maximum stress due to **bending** in the *sweep arm*. Estimate the tension in the belt (not shown) that would be placed over the *driving pulley*.

This style of horse power is very convenient and popular, because, owing to its construction, it has many advantages not found in down powers. It is especially adapted for use in a barn where several horses are kept, or in small livery stables. The power can be bolted to the timbers above the driveway and machines can be set on the floor either above or below the power. When not in use the center post can be lifted from its socket and put out of the way, leaving the floor clear for other purposes. Then when power is used again all that is necessary is to set the post in place, hitch the horse to the sweep and go ahead. The center post which we furnish is made of 6-inch by 6-inch timber and is 12 feet long. It is amply strong and can be cut to any desired length. The 1 1/8-inch driving shaft, to which the pulley is attached, is regularly made so that the measurement from center of master wheel to center of pulley face is 3 feet 3 inches, but additional shafting can be coupled to this shaft so as to change position of pulley or allow the use of other pulleys. The driving pulley is 18 inches in diameter with 3-inch face and makes 37 1/3 revolutions to one round of the horse, or about 135 revolutions per minute. Additional shafting or change in size of pulley is extra. Length of sweep from center post to eye-bolt is 7 feet 6 inches. For driving small feed cutters, corn shellers, feed grinders, wood saws, etc., this power cannot be excelled. Weight, 450 pounds. Shipped direct from factory in Southeastern Wisconsin.   No.   32R1823

....$18.45[2]

---

2. 1902 Edition of The Sears, Roebuck Catalogue.

8.3   A thin Aluminum tube, whose wall thickness is 1 *mm*, carries a torque 80% of the torque required for the onset of yield. The radius of the tube is 20 *mm*.

Show that an estimation of shear stress, based upon the assumption that it is uniformly distributed over the thickness of the tube (and using an estimate of *J* that is linear in the wall thickness), gives a value within 10% of that computed using the full expressions for shear stress and polar moment of inertia.

8.4   A solid aluminum, circular shaft has length 0.25 *m* and diameter 5 *mm*. How much does one end rotate relative to the other if a torque about the shaft axis of 10 *N-m* is applied?

8.5   A relatively thin walled tube and a solid circular shaft have the same cross-sectional area. You are to compare the torsional stiffness of one to the other.

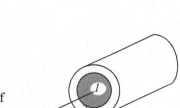

$2\pi Rt \cong A = \pi r^2$

4a) What does the phrase "torsional stiffness" mean?

4b) What is the ratio of the torsional stiffness of the tube to that of the solid shaft?

8.6   A composite, cylindrical shaft has a core of one material, #1, bonding firmly to an outer, concentric, hollow shaft of another material, #2. It can be shown that the shear strain at any radial distance *r* from the center of the shaft is still given by

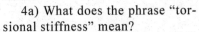

$$\gamma = r \cdot \frac{d\phi}{dz}$$

ie., as if it were a solid, homogeneous shaft.

Let R_1 be the radius of the core, R_2 that of the hollow shaft, take $G_2 = G$ and $G_1 = 2G$ respectively be the shear modulus of the two materials, construct an expression for the shear stress distribution as a function of r in terms of the applied torque and other relevant properties.

9

Stresses: Beams in Bending

The organization of this chapter mimics that of the last chapter on torsion of circular shafts but the story about stresses in beams is longer, covers more territory, and is a bit more complex. In torsion of a circular shaft, the action was all shear; contiguous cross sections sheared over one another in their rotation about the axis of the shaft. Here, the major stresses induced due to bending are normal stresses of tension and compression. But the state of stress within the beam includes *shear stresses due to the shear force* in addition to the major *normal stresses due to bending* although the former are generally of *smaller order* when compared to the latter. Still, in some contexts shear components of stress must be considered if failure is to be avoided.

Our study of the deflections of a shaft in torsion produced a relationship between the applied torque and the angular rotation of one end of the shaft about its longitudinal axis relative to the other end of the shaft. This had the form of a stiffness equation for a linear spring, or truss member loaded in tension, i.e.,

$$M_T = (GJ/L) \cdot \phi \qquad \text{is like} \qquad F = (AE/L) \cdot \delta$$

Similarly, the rate of rotation of circular cross sections was a constant along the shaft just as the *rate of displacement* if you like, $\dfrac{\partial u}{\partial x}$, the extensional strain was constant along the truss member loaded solely at its ends.

We will construct a similar relationship between the moment and the radius of curvature of the beam in bending as a step along the path to fixing the normal stress distribution. We must go further if we wish to determine the transverse displacement and slope of the beam's longitudinal axis. The deflected shape will generally vary as we move along the axis of the beam, and how it varies will depend upon how the loading is distributed over the span. Note that we could have considered a *torque per unit length* distributed over the shaft in torsion and made our life more complex – the rate of rotation, the $d\phi/dz$ would then not be constant along the shaft.

In the next chapter, we derive and solve a differential equation for the transverse displacement as a function of position along the beam. Our exploration of the behavior of beams will include a look at how they might *buckle*. Buckling is a mode of failure that can occur when member loads are well below the yield or fracture strength. Our prediction of *critical buckling loads* will again come from a study of the deflections of the beam, but now we must consider the possibility of *relatively large deflections*.

In this chapter we construct relations for the normal and shear stress components at any point within the the beam's cross-section. To do so, to resolve the indeterminacy we confronted back in chapter 3, we must first consider the deformation of the beam.

9.1   Compatibility of Deformation

We consider first the deformations and displacements of a beam in *pure bending*. **Pure bending is said to take place over a finite portion of a span when the bending moment is a constant over that portion.** Alternatively, a portion of a beam is said to be in a state of pure bending when the shear force over that portion is zero. The equivalence of these two statements is embodied in the differential equilibrium relationship

$$\frac{dM_b}{dx} = -V$$

Using symmetry arguments, we will be able to construct a displacement field from which we deduce a compatible state of strain at every point within the beam. The constitutive relations then give us a corresponding stress state. With this in hand we pick up where we left off in section 3.2 and relate the displacement field to the (constant) bending moment requiring that the stress distribution over a cross section be equivalent to the bending moment. This produces a *moment-curvature relationship*, a stiffness relationship which, when we move to the more general case of varying bending moment, can be read as a differential equation for the transverse displacement.

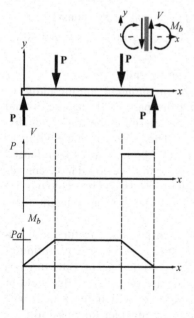

We have already worked up a *pure bending* problem; the *four point bending* of the simply supported beam in an earlier chapter. Over the midspan, $L/4 < x < 3L/4$, the bending moment is constant, the shear force is zero, the beam is in pure bending.

We cut out a section of the beam and consider how it might deform. In this, we take it as given that we have a beam showing a cross section symmetric with respect to the plane defined by $z=0$ and whose shape does not change as we move along the span. We will claim, on the basis of symmetry that **for a beam in pure bending, plane cross sections remain plane and perpendicular to the lon-**

gitudinal axis. For example, postulate that the cross section *CD* on the right does not remain plane but bulges out.

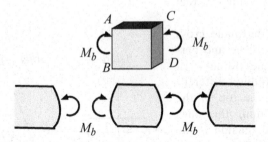

Now run around to the other side of the page and look at the section *AB*. The moment looks the same so section *AB* too must bulge out. Now come back and consider the portion of the beam to the right of section *CD*; its cross section too would be bulged out. But then we could not put the section back without gaps along the beam. This is an incompatible state of deformation.

Any other deformation *out of plane*, for example, if the top half of the section dished in while the bottom half bulged out, can be shown to be incompatible with our requirement that the beam remain all together in one continuous piece. **Plane cross sections must remain plane.**

That cross sections remain perpendicular to the longitudinal axis of the beam follows again from symmetry – demanding that the same cause produces the same effect.

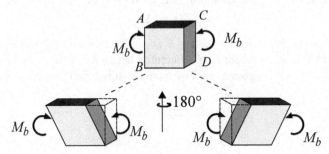

The two alternative deformation patterns shown above are equally plausible – there is no reason why one should occur rather than the other. But rotating either about the vertical axis shown by 180 degrees produces a contradiction. Hence they are both impossible. **Plane cross sections remain perpendicular to the longitudinal axis of the beam.**

The deformation pattern of a differential element of a beam in pure bending below is the one that prevails.

Here we show the plane cross sections remaining plane and perpendicular to the longitudinal axis. We show the longitudinal differential elements near the top of the beam in compression, the ones near the bottom in tension – the anticipated effect of a positive bending moment M_b, the kind shown. We expect then that there is some longitudinal axis which is neither compressed nor extended,

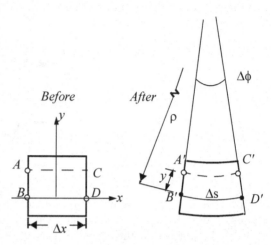

an axis[1] which experiences no change in length. We call this particular longitudinal axis the *neutral axis*. We have positioned our x,y reference coordinate frame with the x axis coincident with this neutral axis.

We first define a *radius of curvature* of the deformed beam in pure bending.

Because plane cross sections remain plane and perpendicular to the longitudinal axes of the beam, the latter deform into arcs of concentric circles. We let the radius of the circle taken up by the neutral axis be ρ and since the differential element of its length, BD has not changed, that is $BD = B'D'$, we have

$$\rho \cdot \Delta\phi = \Delta s = \Delta x$$

where $\Delta\phi$ is the angle subtended by the arc $B'D'$, Δs is a differential element along the deformed arc, and Δx, the corresponding differential length along the undeformed neutral axis. In the limit, as Δx or Δs goes to zero we have, what is strictly a matter of geometry

$$\frac{d\phi}{ds} = 1/\rho$$

where ρ is the *radius of curvature* of the neutral axis.

Now we turn to the extension and contraction of a longitudinal differential line element lying off the neutral axis, say the element AC. Its extensional strain is defined by

$$\varepsilon_x(y) = \lim_{\Delta s \to 0} (A'C - AC)/AC$$

---

1.   We should say "plane", or better yet, "surface" rather than "axis" since the beam has a depth, into the page.

Now AC, the length of the differential line element in its undeformed state, is the same as the length BD, namely $AC = BD = \Delta x = \Delta s$ while its length in the deformed state is $A'C' = (\rho - y) \cdot \Delta\phi$

where y is the vertical distance from the neutral axis.

We have then, using the fact that $\rho\Delta\phi = \Delta s$

we obtain
$$\varepsilon_x(y) = \lim_{\Delta s \to 0}\left[\frac{(\rho - y)\cdot\Delta\phi - \Delta s}{\Delta s}\right] = \lim_{\Delta s \to 0} -(y\Delta\phi/\Delta s)$$

or, finally,

$$\varepsilon_x(y) = -y\cdot\left(\frac{d\phi}{ds}\right) = -(y/\rho)$$

We see that the strain varies linearly with y; the elements at the top of the beam are in compression, those below the neutral axis in extension. Again, this assumes a positive bending moment M_b. It also assumes that there is no other cause that might engender an extension or contraction of longitudinal elements such as an axial force within the beam. If the latter were present, we would *superimpose* a uniform extension or contraction on each longitudinal element.

The extensional strain of the longitudinal elements of the beam is the most important strain component in pure bending. The shear strain γ_{xy} we have shown to be zero; right angles formed by the intersection of cross sectional planes with longitudinal elements remain right angles. This too is an important result. Symmetry arguments can also be constructed to show that the shear strain component γ_{xz} is zero for our span in pure bending, of uniform cross section which is symmetric with respect to the plane defined by the locus of all points $z=0$.

In our discussion of strain-displacement relationships, you will find a displacement field defined by $u(x,y) = -\kappa\,xy$; $v(x,y) = \kappa\,x^2/2$ which yields a strain state consistent in most respects with the above. In our analysis of pure bending we have not ruled out an extensional strain in the y direction which this displacement field does. In the figure we show the deformed configuration of, what we can now interpret as, a short segment of span of the beam.

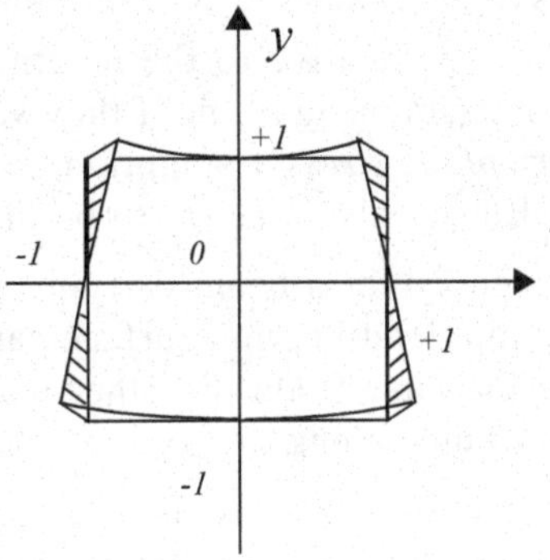

If the κ is interpreted as the reciprocal of the radius of curvature of the neutral axis, the expression for the extensional strain ε_x derived in an earlier chapter is totally in accord with what we have constructed here. κ is called the *curvature* while, again, ρ is called the *radius of curvature*.

Summarizing, we have, for *pure bending* — the case when the bending moment is constant over the whole or some section of a beam — that plane cross-sections remain plane and perpendicular to the neutral axis, (or surface), that the neutral axis deforms into the arc of a circle of radius of curvature, ρ, and longitudinal elements experience an extensional strain ε where:

$$\frac{d\phi}{ds} = 1/\rho$$

$$and$$

$$\varepsilon_x(y) = -(y/\rho)$$

9.2 Constitutive Relations

The stress-strain relations take the form

$$\varepsilon_x = (1/E) \cdot [\sigma_x - \nu(\sigma_y + \sigma_z)] \qquad\qquad 0 = \sigma_{xy}/G$$
$$\varepsilon_y = (1/E) \cdot [\sigma_y - \nu(\sigma_x + \sigma_z)] \qquad\qquad 0 = \sigma_{xz}/G$$
$$\varepsilon_z = (1/E) \cdot [\sigma_z - \nu(\sigma_x + \sigma_y)] \qquad\qquad \gamma_{yz} = \sigma_{yz}/G$$

We now assume that the stress components σ_z and σ_{yz} (hence γ_{yz}), can be neglected, taken as zero, arguing that for beams whose cross section dimensions are small relative to the length of the span[2], these stresses can not build to any appreciable magnitude if they vanish on the surface of the beam. This is the ordinary *plane stress* assumption.

But we also take σ_y, to be insignificant, as zero. This is a bit harder to justify, especially for a beam carrying a distributed load. In the latter case, the stress at the top, load-bearing surface cannot possibly be zero but must be proportional to the load itself. On the other hand, on the surface below, (we assume the load is distributed along the top of the beam), is *stress free* so σ_y, must vanish there. For the moment we make the assumption that it is negligible. When we are through we will compare its possible magnitude to the magnitude of the other stress components which exist within, and vary throughout the beam.

---

2. Indeed, this may be taken as a geometric attribute of what we allow to be called a beam in the first place.

With this, our stress-strain relations reduce to three equations for the normal strain components in terms of the only significant stress component σ_x. The one involving the extension and contraction of the longitudinal fibers may be written

$$\sigma_x(y) = E \cdot \varepsilon_x = -y \cdot (E/\rho)$$

The other two may be taken as machinery to compute the extensional strains in the y,z directions, once we have found σ_x.

9.3 The Moment/Curvature Relation

The figure below shows the stress component $\sigma_x(y)$ distributed over the cross-section. It is a linear distribution of the same form as that considered back in an earlier chapter where we toyed with possible stress distributions which would be equivalent to a system of zero resultant force and a couple.

But now we know for sure, for compatible deformation in pure bending, the exact form of how **the normal stress** must vary over the cross section. According to derived expression for the strain, ε_x, σ_x **must be a linear distribution** in y.

How this *normal stress due to bending* varies with x, the position along the *span* of the beam, depends upon how the *curvature*, $1/\rho$, varies as we move along the beam. For the

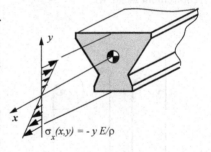

$\sigma_x(x,y) = -y\,E/\rho$

case of pure bending, out analysis of compatible deformations tells us that the curvature is constant so that $\sigma_x(x,y)$ does not vary with x and we can write $\sigma_x(x,y) = \sigma_x(y)$, a (linear) function of y alone. This is what we would expect since the bending moment is obtained by integration of the stress distribution over the cross section: if the bending moment is constant with x, then σ_x should be too. We show this in what follows.

To relate the bending moment to the curvature, and hence to the stress σ_x, we repeat what we did in an earlier exploration of possible stress distributions within beams, first determining the consequences of our requirement that **the resultant force in the axial direction be zero, i.e.,**

$$\int_{Area} \sigma_x \cdot dA = -(E/\rho) \cdot \int_{Area} y \cdot dA = 0 \quad \text{so} \quad \boxed{\int_{Area} y \cdot dA = 0}$$

But what does that tell us? It tells us that **the neutral axis, the longitudinal axis that experiences no extension or contraction, passes through the centroid of the cross section of the beam**. Without this requirement we would be left floating in space, not knowing from whence to measure y. The centroid of the cross section is indicated on the figure.

That this is so, that is, the requirement that our reference axis pass through the centroid of the cross section, follows from the definition of the location of the centroid, namely

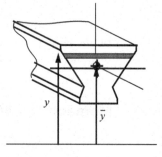

$$\bar{y} \equiv \frac{\int_A y \cdot dA}{A}$$

If y is measured relative to the axis passing thru the centroid, then $\bar{y}$ is zero, our requirement is satisfied.

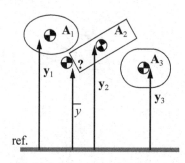

If our cross section can be viewed as a composite, made up of segments whose centroids are easily determined, then we can use the definition of the centroid of a single area to obtain the location of the centroid of the composite as follows.

Consider a more general collection of segments whose centroid locations are known relative to some reference: From the definition of the location of the centroid we can write

$$\bar{y} = \frac{\int_{A_1} y \cdot dA_1 + \int_{A_2} y \cdot dA_2 + \int_{A_3} y \cdot dA_3}{A_1 + A_2 + A_3} = \frac{\bar{y}_1 \cdot A_1 + \bar{y}_2 \cdot A_2 + \bar{y}_3 \cdot A_3}{A}$$

where A is the sum of the areas of the segments.
We use this in an exercise, to wit:

Exercise 9.1

Determine the location of the neutral axis for the "T" cross-section shown.

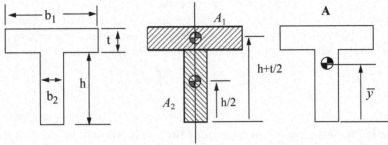

We seek the centroid of the cross-section. Now, because the cross-section is symmetric with respect to a vertical plane perpendicular to the page and bisecting the top and the bottom rectangles, the centroid must lie in this plane or, since this plane appears as a line, it must lie along the vertical line, AA'. To find where along this vertical line the centroid is located, we first set a reference axis for measuring vertical distances. This could be chosen anywhere; I choose to set it at the base of the section.

I let $\bar{y}$ be the distance from this reference to the centroid, yet unknown. From the definition of the location of the centroid and its expression at the top of this page in terms of the centroids of the two segments, we have

$$\bar{y} \cdot A = (h/2) \cdot A_2 + (h + t/2) \cdot A_1$$

where

$$A = h \cdot b_2 + t \cdot b_1, \qquad A_2 = h \cdot b_2 \qquad \text{and} \qquad A_1 = t \cdot b_1$$

This is readily solved for $\bar{y}$ given the dimensions of the cross-section; the centroid is indicated on the figure at the far right.

Taking stock at this point, we see that, for the case of "pure bending", we have that the normal strain and normal stress vary linearly over the cross-section. If we measure y from the centroid of the section we have

$$\sigma_x(y) = E \cdot \varepsilon_x = -y \cdot (E/\rho)$$

In this case, the resultant force due to σ_x is zero.

The resultant moment due to σ_x must equal the bending moment. In fact, we have:

$$-M_b = \int_{Area} y \cdot [\sigma_x dA]$$

This relationship ties the bending moment to the curvature.

Within the integrand, the term within the brackets is a differential element of force due to the stress distributed over the differential element of area $b(y)dy$. The negative sign in front of M_b is necessary because, if σ_x were positive at y positive, on a positive x face, then the differential element of force $\sigma_x\, dA$ would produce a moment about the **negative** z axis, which, according to our convention for bending moment, would be a **negative** bending moment. Now substituting our known linear distribution for the stress σ_x, we obtain

$$M_b = (E/\rho) \cdot \int_{Area} y^2 \cdot dA$$

The integral is again just a function of the geometry of the cross section. It is often called **the moment of inertia about the z axis**. We will label it I. That is[3]

$$\boxed{I = \int_{Area} y^2 \cdot dA}$$

Our *Moment/Curvature relationship* is then:

$$\boxed{\begin{array}{c} M_b = (EI) \cdot \dfrac{1}{\rho} \\[2mm] \text{where the curvature is also defined by} \\[2mm] \kappa = \dfrac{1}{\rho} = \dfrac{d\phi}{ds} \end{array}}$$

Here is a most significant result, very much of the same form of the stiffness relation between the torque applied to a shaft and the rate of twist — but with a quite different ϕ

$$M_T = (GJ) \cdot \frac{d\phi}{dz}$$

and of the same form of the stiffness relation for a rod in tension

$$F = (AE) \cdot \frac{du}{dx}$$

This moment-curvature relationship tells us the radius of curvature of an initially straight, uniform beam of symmetric cross-section, when a bending moment M_b is applied. And, in a fashion analogous to our work a circular shaft in torsion, we can go back and construct, using the moment curvature relation, an expression for the normal stress in terms of the applied bending moment. We obtain

---

3. Often subscripts are added to I, e.g., I_{yy} or I_z; both are equally acceptable and/or confusing. The first indicates the integral is over y and y^2 appears in the integrand; the second indicates that the moment of inertia is "about the z axis", as if the plane area were rotating (and had inertia).

$$\sigma_x(y) = -\frac{M_b(x) \cdot y}{I}$$

Note again the similarity of form with the result obtain for the shaft in torsion, $\tau = M_T r/J$, (but note here the negative sign in front) and observe that the maximum stress is going to occur either at the top or bottom of the beam, whichever is further off the neutral axis (just as the maximum shear stress in torsion occurs at the outermost radius of the shaft).

Here we have revised Galileo. We have answered the question he originally posed. While we have done so strictly for the case of "pure bending", this is no serious limitation. In fact, we take the above relationships to be accurate enough for the design and analysis of most beam structures even when the loading is not pure bending, even when there is a shear force present.

It remains to develop some machinery for the calculation of the moment of inertia, I, when the section can be viewed as a composite of segments whose "local" moments of inertia are known. That is, we need *a parallel axis theorem* for evaluating the moment of inertia of a cross-section.

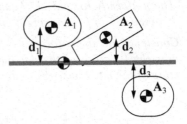

We use the same composite section as above and seek the total moment of inertia of all segments with respect to the centroid of the composite. We first write I as the sum of the I's

$$I = \int_{Area} y^2 \cdot dA = \int_{A_1} y^2 \cdot dA_1 + \int_{A_2} y^2 \cdot dA_2 + \int_{A_3} y^2 \cdot dA_3$$

then, for each segment, express y in terms of d and a local variable of integration, η.

That is, we let $y = d_1 + \eta_1$ etc. and so obtain;

$$I = \int_{A_1} (d_1 + \eta_1)^2 \cdot dA_1 + \text{etc.} = \int_{A_1} (d_1^2 + 2d_1\eta_1 + \eta_1^2) \cdot dA_1$$

$$= d_1^2 A_1 + \int_{A_1} 2d_1\eta_1 \cdot dA_1 + \int_{A_1} \eta_1^2 \cdot dA_1 + \text{etc.}$$

Now the middle term in the sum on the right vanishes since $\int_{A_1} 2d_1\eta_1 \cdot dA_1 = 2d_1 \int_{A_1} \eta_1 \cdot dA_1$ and η_1 is measured from the local centroid. Furthermore, the last term in the sum on the right is just the local moment of inertia. The end result is the parallel axis theorem (employed three times as indicated by the "etc".)

$$I = d_1^2 A_1 + I_1 \quad + \text{etc.}$$

The bottom line is this: Knowing the local moment of inertia (with respect to the centroid of a segment) we can find the moment of inertia with respect to any axis parallel to that passing through the centroid of the segment by adding a term equal to the product of the area and the square of the distance from the centroid to the arbitrarily located, parallel axis. Note that the moment of inertia is a minimum when taken with respect to the centroid of the segment.

Exercise 9.2

The uniformly loaded "I" beam is simply supported at its left end and at a distance L/5 in from its right end.

Construct the shear-force and bending-moment diagrams, noting in particular the location of the maximum bending moment. Then develop an estimate of the maximum stress due to bending.

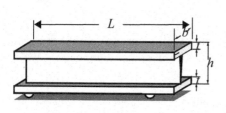
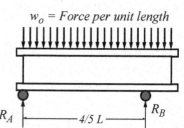

An isolation of the entire span and requiring equilibrium gives the two vertical reactions,

$$R_A = (3/8)w_o L \qquad and \qquad R_B = (5/8)w_o L$$

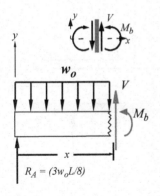

A section of the beam cut between the two supports will enable the evaluation of the shear force and bending moment within this region. We have indicated our positive sign convention as the usual. Force equilibrium gives the shear force

$$V(x) = -(3/8)w_o L + w_o x$$

which, we note, passes through zero at $x = (3/8)L$ and, as we approach the right support, approaches the value $(17/40)w_o L$. At this point, the shear force suffers a discontinuity, a jump equal in magnitude to the reaction at B.   Its value just to the right of the right support is then $-(1/5)w_o L$. Finally, it moves to zero at the end of the beam at the same rate, w_0, as required by the differential equilibrium relation

$$\frac{dV}{dx} = w_o$$

We could, if we wish, at this point use the same free body diagram above to obtain an expression for the bending moment distribution. I will not do this. Rather I will construct the bending moment distribution using insights gained from evaluating M_b at certain critical points and reading out the implications of the other differential, equilibrium relationship,

$$\frac{dM_b}{dx} = -V(x)$$

One interesting point is at the left end of the beam. Here the bending moment must be zero since the roller will not support a couple. Another critical point is at $x=(3/8)L$ where the shear force passes through zero. Here the **slope** of the bending moment must vanish. We can also infer from this differential, equilibrium relationship that the slope of the bending moment distribution is positive to the left of this point, negative to the right. Hence at this point the bending moment shows a local maximum. We *cannot* claim that it is the maximum over the whole span until we check all boundary points and points of shear discontinuity.

Furthermore, since the shear force is a linear function of x, the bending moment must be quadratic in this region, symmetric about $x=(3/8)L$. Now since the distance from the locus of the local maximum to the roller support at the right is greater than the distance to the left end, the bending moment will diminish to less than zero at the right support. We can evaluate its magnitude by constructing an isolation that includes the portion of the beam to the **right** of the support.

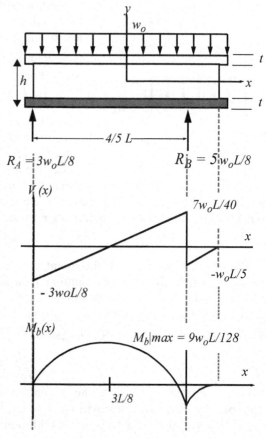

We find, in this way that, at $x=4L/5$, $M_b = - w_o L^2/50$.

At the right support there is a discontinuity in the slope of the bending moment equal to the discontinuity in the value of the shear force. The jump is just equal to the reaction force R_B. In fact the slope of the bending moment must switch from negative to positive at this point because the shear force has changed sign. The character of the bending moment distribution from the right support point out to the right end of the beam is fully revealed noting that, first, the bending moment must go to zero at the right end, and, second, that since the shear force goes to zero there, so must the slope of the bending moment.

All of this enables sketching the shear force and bending moment distributions shown. We can now state definitively that the maximum bending moment occurs

at $x= 3L/8$. Its value is $M_b\big|_{max} = (9/128)w_o L$

The maximum stress due to this maximum bending moment is obtained from

$$\sigma_x\big|_{max} = -\frac{y \cdot (M_b\big|_{max})}{I}$$

It will occur at the top and bottom of the beam where $y = \pm h/2$, measured from the neutral, centroidal axis, attains its maximum magnitude. At the top, the stress will be compressive while at the bottom it will be tensile since the maximum bending moment is a positive quantity.

We must still evaluate the moment of inertia I. Here we will estimate this quantity assuming that $t < h$, *and/or b*, that is, the *web of the I beam* is thin, or has negligible area relative to the *flanges* at the top and bottom. Our estimate is then

$$I \approx 2 \cdot [(h/2)^2] \cdot [t \cdot b]$$

The last bracketed factor is the area of one flange. The first bracketed factor is the square of the distance from the y origin on the neutral axis to the centroid of the flange, or an estimate thereof. The factor of two out front is there because the two flanges contribute equally to the moment of inertia. How good an estimate this is remains to be tested. With all of this, our estimate of the maximum stress due to bending is

$$|\sigma_x|_{max} \approx (9/128) \cdot \frac{w_o L^2}{(thb)}$$

9.4 Shear Stresses in Beams

In this last exercise we went right ahead and used an equation for the normal stress due to bending constructed on the assumption of a particular kind of loading, namely, *pure bending*, a loading which produces no shear force within the beam. Clearly we are not justified in this assumption when a distributed load acts over the span. No problem. The effect of the shear force on the normal stress distribution we have obtained is negligible. Furthermore, the effect of a shear force on the deflection of the beam is also small. All of this can be shown to be accurate enough for most engineering work, at least for a true beam, that is when its length is much greater than any of its cross-section's dimensions. We will first show that the shear stresses due to a shear force are small with respect to the normal stresses due to bending.

Reconsider Galileo's end loaded, cantilever beam. At any section, x, a shear force, equal to the end load, which we now call P, acts in accord with the requirement of static equilibrium. I have shown the end load as acting up. The shear force is then a positive quantity according to our convention.

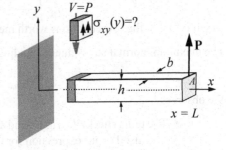

We postulate that this shear force is distributed over the plane cross section at x in the form of a shear stress σ_{xy}. Of course there is a normal stress σ_x distributed over this section too, with a resultant moment equal to the bending moment at the section. But we do not show that on our picture just yet.

What can we say about the shear stress distribution $\sigma_{xy}(y)$? For starters we can claim that it is only a function of y, not of z as we have indicated *(nor of x in this case since $V(x)$ is constant)*. The truth of this claim depends upon the shape of the contour of the cross-section as we shall see. For a rectangular cross section it's a valid claim.

We can also claim with more assurance that the shear stress must vanish at the top and bottom of the beam because we know from chapter 4 that at every point we have $\sigma_{xy} = \sigma_{yx}$; but at the top and bottom surfaces σ_{yx} vanishes (there is no applied force in the horizontal direction) so σ_{xy} must vanish at $y=\pm\ h/2$.

We expect the shear stress to grow continuously to some finite value at some point in the interior. We expect, for a continuous, homogeneous material, that it will vary smoothly with y. Its maximum value, if this is the case, ought not to be too different from its **mean** value defined by

$$\sigma_{xy}\big|_{mean} \approx \frac{V}{A} = \frac{P}{(bh)}$$

Now compare this with the maximum of the normal stress due to bending. Recalling that the maximum bending moment is PL at $x=0$, at the wall, and using our equation for pure bending, we find that

$$\sigma_x\big|_{max} = \frac{M_b \cdot y}{I} = \frac{PL(h/2)}{I}$$

I now evaluate the moment of inertia of the cross-section, I have

$$I = \int_{Area} y^2 \cdot dA = \int_{-h/2} y^2 \cdot dA$$

This yields, with careful attention to the limits of integration,

$$\boxed{I = \frac{bh^3}{12}}$$

which is one of the few equations worth memorizing in this course.[4]

The maximum normal stress due to bending is then

$$\sigma_x\big|_{max} = 6\frac{PL}{(bh^2)}$$

We observe:

- The units check; the right hand side has dimensions of stress, F/L^2. This is true also for our expression for the average shear stress.

- The ratio of the maximum shear stress to the maximum normal stress due to bending is *on the order of*

$$\sigma_{xy}\big|_{max} / \sigma_x\big|_{max} = \text{Order (h/L)}$$

---

4.   Most practitioners say this as "*bee* h cubed over twelve". Like "sigma is equal to *em y* over *eye*" it has a certain ring to it.

which if the beam is truly a beam, is on the order of 0.1 or 0.01 — as Galileo anticipated!

- While the shear stress is small relative to the normal stress due to bending, it does not necessarily follow that we can neglect it even when the ratio of a dimension of the cross section to the length is small. In particular, in built up, or *composite* beams, excessive shear can be a cause for failure.

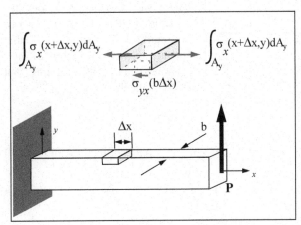

We next develop a more accurate, more detailed, picture of the shear stress distribution making use of an ingeneous free-body diagram. Look left.

We show the forces acting on a differential element of the cantilever, of length Δx cut from the beam at some y station which is arbitrary. (We do not show the shear stress σ_{xy} acting on the two

"x faces" of the element as these will not enter into our analysis of force equilibrium in the x direction).

For force equilibrium in the x direction, we must have

$$\int_{A_y}\sigma_x(x + \Delta x, y)dA - \int_{A_y}\sigma_x(x, y)dA = \sigma_{yx}(y)b\Delta x$$

This can be written

$$\int_{A_y}\frac{\sigma_x(x + \Delta x, y) - \sigma_x(x, y)}{\Delta x} \cdot dA = b\sigma_{yx}(y)$$

Which, as Δx approaches zero, yields $\qquad \int_{A_y}\frac{\partial}{\partial x}\sigma_x(x, y) \cdot dA = b\sigma_{yx}(y)$

Now, our engineering beam theory says

$$\sigma_x(x, y) = -\frac{M_b(x) \cdot y}{I} \qquad \text{and we have from before} \qquad \frac{d}{dx}M_b(x) = -V$$

so our equilibrium of forces in the direction of the longitudinal axis of the beam, on an oddly chosen, section of the beam (of length Δx and running from y up to the top of the beam) gives us the following expression for the shear stress σ_{yx} and thus σ_{xy} namely:

$$\sigma_{yx}(y) = \sigma_{xy}(y) = \frac{V}{bI} \cdot \int_{A_y} y \cdot dA$$

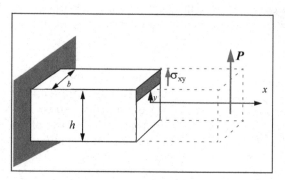

For a rectangular section, the element of area can be written

$$dA = b \cdot d\eta$$

where we introduce η as our "y" variable of integration so that we do not confuse it with the "y" that appears in the lower limit of integration. We have then, noting that the b's cancel:

$$\sigma_{xy}(y) = \frac{V}{I} \int_{y}^{h/2} \eta \cdot d\eta$$

which, when integrated gives $\sigma_{xy}(y) = \frac{V}{2I} \cdot \left[\left(\frac{h}{2}\right)^2 - y^2\right]$ i.e., a parabolic distribution with maximum at y=0.

The maximum value is, once putting $I = bh^3/12$, $\sigma_{xy}(y) = \frac{3}{2} \cdot \frac{P}{bh}$ where we have assumed an end-loaded cantilever as in the figure.

This is to be compared with the average value obtained in our order of magnitude analysis. The order of magnitude remains essentially less than the maximum normal stress due to bending by a factor of (h/L).

9.5   Stresses within a Composite Beam

A *composite* beam is composed of two or more elemental structural forms, or different materials, bonded, knitted, or otherwise joined together. *Composite materials or forms* include such heavy handed stuff as concrete (one material) reinforced with steel bars (another material); high-tech developments such as tubes built up of graphite fibers embedded in an epoxy matrix; sports structures like *laminated* skis, the poles for vaulting, even a golf ball can be viewed as a *filament wound* structure encased within another material. *Honeycomb* is another example of a composite – a *core material*, generally light-weight and relatively flimsy, maintains the distance between two *face sheets*, which are relatively sturdy with respect to *in-plane* extension and contraction.

To determine the moment/curvature relation, the normal stresses due to bending, and the shear stresses within a composite beam, we proceed through the *pure bending* analysis all over again, making careful note of when we must alter our constructions due to the *inhomogeneity* of the material.

Compatibility of Deformation

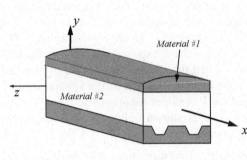

Our analysis of deformation of a beam in pure bending included no reference to the material properties or how they varied throughout the beam. We did insist that the cross-section be symmetric with respect to the $z=0$ plane and that the beam be uniform, that is, no variation of geometry or properties as we move in the longitudinal direction. A composite structure of the kind shown below would satisfy these conditions.

Constitutive Relations

We have two materials so we must necessarily contend with two sets of material properties. We still retain the assumptions regarding the smallness of the stress components σ_y, σ_z and τ_{yz} in writing out the relations for each material. For material #1 we have $\sigma_x = -E_1 \cdot (y/\rho)$ while for material #2 $\sigma_x = -E_2 \cdot (y/\rho)$

Equilibrium

Equivalence of this normal stress distribution sketched below to zero resultant force and a couple equal to the bending moment at any station along the span proceeds as follows:

For zero resultant force we must have

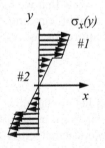

$$\int\limits_{Area_1} \sigma_x \cdot dA_1 + \int\limits_{Area_2} \sigma_x \cdot dA_2 = 0$$

Upon substituting our strain-compatible variation of stress as a function of y into this we obtain, noting that the radius of curvature, ρ is a common factor,

$$E_1 \int\limits_{Area_1} y \cdot dA_1 + E_2 \int\limits_{Area_2} y \cdot dA_2 = 0$$

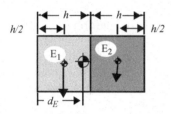

What does this mean?   Think of it as a machine for computing the location of the unstrained, neutral axis, $y = 0$. However, in this case it is located, not at the centroid of the cross-sectional area, but at the *centroid of area weighted by the elastic moduli*. The meaning of this is best exposed via a short thought experiment.   Turn the composite section over on its side. For ease of visualization of the special effect I want to induce I consider a composite cross section of two rectangular subsections **of equal area** as shown below. Now think of the elastic modulus as a weight density, and assume $E_1 > E_2$, say $E_1 = 4 E_2$.

This last equation is then synonymous with the requirement that the location of the neutral axis is at the center of gravity of the elastic-modulus-as-weight-density configuration shown.[5] Taking moments about the left end of the tipped over cross section we must have

$$[AE_1 + AE_2] \cdot d_E = (AE_1) \cdot (h/2) + (AE_2) \cdot (3h/2)$$

With $E_1 = 4 E_2$ this gives $d_E = (7/10) \cdot h$ for the location of the *E-weighted centroid*

Note that if the elastic moduli were the same the centroid would be at h, at the mid-point. On the other hand, if the elastic modulus of material #2 were greater than that of material #1 the centroid would shift to the right of the interface between the two.

Now that we have a way to locate our neutral axis, we can proceed to develop a moment curvature relationship for the composite beam in pure bending. We require for equivalence

$$-M_b = \int_{Area} y \cdot \sigma_x dA$$

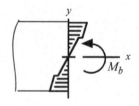

as before, but now, when we replace σ_x with its variation with y we must distinguish between integrations over the two material, cross-sectional areas. We have then, breaking up the area integrals into A_1 over one material's cross section and A_2, the other material's cross section[6]

$$M_b = (E_1/\rho) \cdot \int_{A_1} y^2 \cdot dA_1 + (E_2/\rho) \cdot \int_{A_2} y^2 \cdot dA_2$$

---

5. I have assumed in this sketch that material #1 is *stiffer*, its elastic modulus E_1 is greater than material #2 with elastic modulus E_2.

6. Note that we can have the area of either one or both of the materials distributed in any manner over the cross section in several non-contiguous pieces. Steel reinforced concrete is a good example of this situation. We still, however, insist upon symmetry of the cross section with respect to the *x-y* plane.

The integrals again are just functions of the geometry. I designate them I_1 and I_2 respectively and write

$$M_b/[E_1 I_1 + E_2 I_2] = 1/\rho \quad or \quad M_b/(\overline{EI}) = 1/\rho$$

Here then is our moment curvature relationship for pure bending of a composite beam. It looks just like our result for a homogeneous beam but note

- Plane cross sections remain plane and perpendicular to the longitudinal axis of the beam. Compatibility of Deformation requires this as before.

- The neutral axis is located not at the centroid of area but at the centroid of the E-weighted area of the cross section. In computing the moments of inertia I_1, I_2 the integrations must use y measured from this point.

- The stress distribution is linear within each material but there exists a discontinuity at the interface of different materials. The exercise below illustrates this result. Where the maximum normal stress appears within the cross section depends upon the relative stiffnesses of the materials **as well as** upon the geometry of the cross section.

We will apply the results above to loadings other than pure bending, just as we did with the homogeneous beam. We again make the claim that the effect of shear upon the magnitude of the normal stresses and upon the deflected shape is small although here we are skating on thinner ice – still safe for the most part but thinner. And we will again work up a method for estimating the shear stresses themselves. The following exercise illustrates:

Exercise 9.3

A composite beam is made of a solid polyurethane core and aluminum face sheets. The modulus of elasticity, E for the polyurethane is 1/30 that of aluminum. The beam, of the usual length L, is simply supported at its ends and

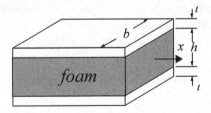

carries a concentrated load P at midspan. If the ratio of the thickness of the aluminum face sheets to the thickness of the core is t/h = 1/20 develop an estimate for the maximum shear stress acting at the interface of the two materials.

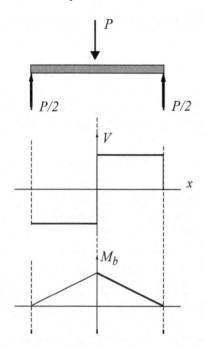

We first sketch the shear force and bending moment diagram, noting that the maximum bending moment occurs at mid-span while the maximum shear force occurs at the ends.

Watch this next totally unmotivated step. I am going to move to estimate the shear stress at the interface of the aluminum and the core. I show an isolation of a differential element of the aluminum face sheet alone. I show the normal stress due to bending and how it varies both over the thickness of the aluminum and as we move from x to $x+\Delta x$. I also show a differential element of a shear force ΔF_{yx} acting on the underside of the differential element of the aluminum face sheet. I **do not** show the shear stresses acting of the x faces; their resultant on the x face is in equilibrium with their resultant on the face $x + \Delta x$.

Equilibrium in the x direction will be satisfied if

$$\Delta F_{yx} = \int_{A_{al}} \Delta \sigma_x dA_{al}$$

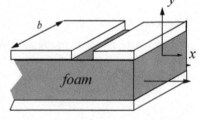

where A_{al} is the cross-sectional area of the aluminum face sheet. Addressing the left hand side, we set

$$\Delta F_{yx} = \sigma_{yx} b \Delta x$$

where σ_{yx} is the shear stress at the interface, the quantity we seek to estimate. Addressing the right hand side, we develop an expression for $\Delta \sigma_x$ using our pure bending result. From the moment curvature relationship for a composite cross section we can write $M_b/(\overline{EI}) = 1/\rho$

The stress distribution within the aluminum face sheet is then[7]

$$\sigma_x = -E_{al} \cdot y/\rho \qquad \text{which is then} \qquad \sigma_x = -E_{al} \cdot y M_b/(\overline{EI})$$

---

7. Note the similarity to our results for the torsion of a composite shaft.

Taking a differential view, as we move a small distance Δx and noting that the only thing that varies with x is M_b, the bending moment, we have

$$\Delta\sigma = - [\, E_{al}\, y\,/\,\overline{(EI)}\,]\Delta M_b(x)$$

But the change in the bending moment is related to the shear force through differential equation of equilibrium which can be written $\Delta M_b = - V(x)\Delta x$. Putting this all together we can write

$$\sigma_{yx} \cdot b\Delta x = \left\{ \int_{A_{al}} E_{al} \cdot V(x) \cdot \frac{y}{\overline{(EI)}} dA_{al} \right\} \cdot \Delta x$$

or, in the limit

$$\sigma_{yx} = \frac{E_{al}(V/b)}{\overline{EI}} \int_{A_{al}} y \cdot dA_{al}$$

This provides an estimate for the shear stress at the interface. Observe:

- This expression needs elaboration. It is essential that you read the phrase $\int_{A_{al}} y\,dA$ correctly. First, y is to be measured from the E-weighted centroid of the cross section (which in this particular problem is at the center of the cross section because the aluminum face sheets are symmetrically disposed at the top and the bottom of the cross section and they are of equal area). Second, the integration is to be performed over the aluminum cross section only. More specifically, from the coordinate $y= h/2$, where one is estimating the shear stress up to the top of the beam, $y = h/2 +t$. This *first moment of area* may be approximated by

$$\int_{A_{al}} y \cdot dA_{al} = bt(h/2)$$

- The shear stress is dependent upon the change in the normal stress component σ_x with respect to x. This resonates with our derivation, back in chapter 3, of the differential equations which ensure equilibrium of a differential element.

- The *equivalent* $\overline{EI}$ can be evaluated noting the relative magnitudes of the elastic moduli and approximating the moment of inertia of the face sheets as $I_{al} = 2(bt)(h/2)^2$ while for the foam we have $I_f = bh^3/12$. This gives

$$\overline{EI} = (5/9)(E_{al} \cdot bth^2)$$

Note the consistent units; FL^2 on both sides of the equation. The foam contributes $1/9^{th}$ to the *equivalent bending stiffness*.

- The magnitude of the shear stress at the interface is then found to be, with V taken as $P/2$ and the first moment of area estimated above,

$$\sigma_{yx}\big|_{interface} = (9/10) \cdot \frac{P}{bh}$$

- The maximum normal stress due to bending will occur in the aluminum[8]. Its value is approximately

$$\sigma_x\big|_{max} = (9/2) \cdot \frac{P}{bh} \cdot \frac{L}{h}$$

which, we note again is on the order of L/h times the shear stress at the interface.

As a further example, we consider a steel-reinforced concrete beam which, for simplicity, we take as a rectangular section.

We assume that the beam will be loaded with a positive bending moment so that the bottom of the beam will be in tension and the top in compression.

We reinforce the bottom with steel rods. They will carry the tensile load. We further assume that the concrete is unable to support any tensile load. So the concrete is only effective in compression over the area of the cross section above the neutral axis.

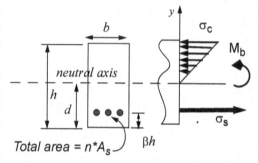

In proceeding, we identify the steel material with material #1 and the concrete with material #2 in our general derivation. We will write

$$E_1 = E_s = 30e06 \text{ psi} \qquad \text{and} \qquad E_2 = E_s = 3.6e06 \text{ psi}$$

The requirement that the resultant force, due to the tensile stress in the steel and the compressive stress in the concrete, vanish then may be written

$$\int_{A_s} \sigma_s \cdot dA_s + \int_{A_c} \sigma_c \cdot dA_c = 0 \quad \text{or} \quad \int_{A_s} -E_s \cdot (y/\rho) \cdot dA_s + \int_{A_c} -E_c \cdot (y/\rho) \cdot dA_c = 0$$

which since the radius of curvature of the neutral axis is a constant relative to the integration over the area, can be written:

$$E_s \int_{A_s} y \cdot dA_s + E_c \int_{A_c} y \cdot dA_c = 0$$

The first integral, assuming that all the steel is concentrated at a distance $(d - \beta h)$ *below* the neutral axis, is just

$$-E_s(d - \beta h)nA_s$$

where the number of reinforcing rods, each of area A_s, is n. The negative sign reflects the fact that the steel lies below the neutral axis.

---

8. The aluminum is stiffer; for comparable extensions, as compatibility of deformation requires, the aluminum will then carry a greater load. But note, the foam may fail at a much lower stress. A separation due to shear at the interface is a possibility too.

The second integral is just the product of the distance to the centroid of the area under compression, $(h-d)/2$, the area $b(h-d)$, and the elastic modulus.

$$E_c \cdot \frac{(h-d)}{2} \cdot b(h-d)$$

The zero resultant force requirement then yields a quadratic equation for d, or d/h, putting it in nondimensional form. In fact

$$\left(\frac{d}{h}\right)^2 - (2+\lambda) \cdot \left(\frac{d}{h}\right) + (1+\beta\lambda) = 0 \qquad \text{where we define} \qquad \lambda = \frac{2E_s n A_s}{E_c bh}$$

This gives $\qquad \frac{d}{h} = \frac{1}{2} \cdot [(2+\lambda) \pm \sqrt{(2+\lambda)^2 - 4(1+\beta\lambda)}]$

There remains the task of determining the stresses in the steel and concrete. For this we need to obtain an expression for the equivalent bending stiffness, $\overline{EI}$.

The contribution of the steel rods is easily obtained, again assuming all the area is concentrated at the distance (d-bh) below the neutral axis. Then

$$I_s = (d - \beta h)^2 n A_s$$

The contribution of the concrete on the other hand, using the transfer theorem for moment of inertia, includes the "local" moment of inertia as well as the transfer term.

$$I_c = b(h-d) \cdot \left(\frac{h-d}{2}\right)^2 + \frac{b \cdot (h-d)^3}{12}$$

Then $\quad \overline{EI} = E_s \cdot I_s + E_c \cdot I_c \quad$ and the stress are determined accordingly, for the steel, by

$$\sigma_s \big|_{tension} = M_b \cdot E_s \cdot \frac{(d - \beta h)}{\overline{EI}}$$

while for the concrete $\qquad \sigma_c \big|_{compression} = M_b \cdot E_c \cdot \frac{y}{\overline{EI}}$

9.6   Problems - Stresses in Beams

9.1   In some of our work we have approximated the moment of inertia of the cross-section effective in bending by

$$I \sim 2 \, (h/2)^2 (bt)$$

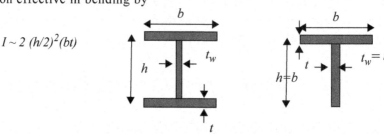

It $t/h \sim 0.01$, or 0.1, estimate the error made by comparing the number obtained from this approximate relationship with the exact value obtained from an integration.

9.2   For a beam with a T section as shown in the problem above:

i) Locate the centroid of the section.

ii) Construct an expression for the moment of inertia about the centroid.

iii) Locate where the maximum tensile stress occurs and express its magnitude in terms of the bending moment and the geometry of the section. Do the same for the maximum compressive stress. In this assume the bending moment puts the top of the beam in compression.

iv) If you take b equal to the h of the I beam, so that the cross-sectional areas are about the same, compare the maximum tensile and compressive stresses within the two sections.

9.3   A steel wire, with a radius of *0.0625 in*, with a yield strength of $120x10^3$ *psi,* is wound around a circular cylinder of radius $R = 20$ *in.* for storage. What if your boss, seeking to save money on storage costs, suggests reducing the radius of the cylinder to $R = 12in$. How do you respond?

9.4   The cross-section of a beam made of three circular rods connected by three thin "shear webs" is shown.

i) Where is the centroid?

ii) What is the moment of inertia of the cross-section?

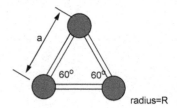

9.5 For the "C" section shown below: Locate the centroid in x and y. Verify

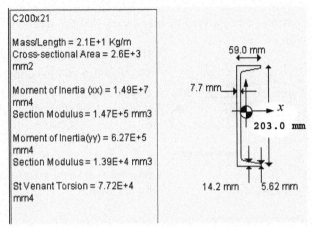

C200x21

Mass/Length = 2.1E+1 Kg/m
Cross-sectional Area = 2.6E+3 mm2

Moment of Inertia (xx) = 1.49E+7 mm4
Section Modulus = 1.47E+5 mm3

Moment of Inertia(yy) = 6.27E+5 mm4
Section Modulus = 1.39E+4 mm3

St Venant Torsion = 7.72E+4 mm4

the values given for the mass/length, the cross-sectional area, and the two moments of inertia. (Note that Moment of Inertia (xx) refers to the moment of inertia about the "x-x" axis, which we have labeled, "I"). That is

$$I = I_{(xx)} = \int_A y^2 \cdot dA$$

The (yy) refers to the moment of inertia about the "y-y" axis.

9.6 A beam is pinned at its left end and supported by a roller at 2/3 the length as shown. The beam carries a uniformly distributed load, w_0, <F/L>

i) Where does the maximum normal stress due to bending occur.

ii) If the beam has an I cross section with

flange width = .5"
section depth = 1.0"
and $t_w = t = 0.121$"

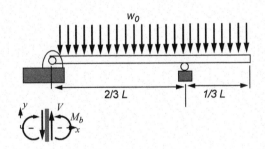

and the length of the beam is 36" and the distributed load is 2 lb/inch, determine the value of the maximum normal stress.

iii) What if the cross section is rectangular of the same height and area? What is the value of the maximum normal stress due to bending?

9.7   A steel reinforced beam is to be made such that the steel and the concrete fail simultaneously.

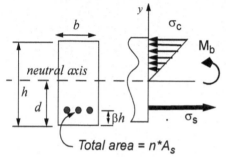

If E_s = 30 e06 psi   steel

E_c= 3.6 e06 psi concrete

and taking

$\sigma_{\text{failure steel}}$ = 40,000 psi

$\sigma_{\text{failure concrete}}$ = 4,000 psi (compression)

how must β be related to d/h for this to be the case?

Now, letting $$\lambda = \frac{2 \cdot E_s \cdot nA_s}{E_c \cdot bh}$$ find d/h and β values for a range of "realistic" values for the area ratio, (nAs/bh), hence for a range of values for Λ.

Make a sketch of one possible composite cross-section showing the location of the reinforcing rod. Take the diameter of the rod as 0.5 inches.

10

Deflections due to Bending

10.1 The Moment/Curvature Relation

Just as we took the *pure bending* construction to be accurate enough to produce useful estimates of the normal stress due to bending for loadings that included shear, so too we will use the same *moment/curvature* relationship to produce a differential equation for the transverse displacement, $v(x)$ of the beam at every point along the neutral axis when the bending moment varies along the beam.

$$\frac{M_b}{EI} = \frac{d\phi}{ds}$$

The moment/curvature relationship itself **is** this differential equation. All we need do is express the curvature of the deformed neutral axis in terms of the transverse displacement. This is a straight forward application of the classical calculus as you have seen perhaps but may also have forgotten. That's ok. For it indeed can be shown that[1]:

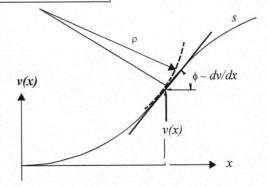

$$\frac{d\phi}{ds} = \frac{\dfrac{d^2 v}{dx^2}}{\left[1 + \left(\dfrac{dv}{dx}\right)^2\right]^{\frac{3}{2}}}$$

---

1. Note in this exact relationship, the independent variable is s, the distance along the curved, deformed neutral, x axis.

There now we have it – once given the bending moment as a function of x all we need do is solve this non-linear, second order, ordinary differential equation for the transverse displacement $v(x)$.

But hold on. When was the last time you solved a second order, **nonlinear** differential equation? Leonhard Euler attacked and resolved this one for some quite sophisticated end-loading conditions back in the eighteenth century but we can get away more cheaply by making our usual assumption of small displacement and rotations.

That is we take $\left(\dfrac{dv}{dx}\right)^2 < 1$ which says that the slope of the deflection is small

with respect to 1.0. Or equivalently that the rotation of the cross section as measured by $\phi \approx (dv/dx)$ is less than 1.0, one radian. In this we note the dimensionless character of the slope. Our moment curvature equation can then be written more simply as

$$\frac{d^2v}{dx^2} = \frac{M_b(x)}{EI}$$

Exercise 10.1

Show that, for the end loaded beam, of length L, simply supported at the left end and at a point L/4 out from there, the tip deflection under the load P is

given by $\Delta = (3/16) \cdot \dfrac{PL^3}{EI}$

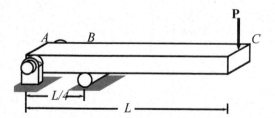

The first thing we must do is determine the bending moment distribution as a function of x. No problem. The system is statically determinate. We first determine the reactions at A and B from an isolation of the whole. We find $R_A = 3\,P$, directed down, and $R_B = 4P$ directed up.

An isolation of a portion to the right of the support at *B* looks very much like Galileo's cantilever. In this region we find a constant shear force equal in magnitude to the end load and a linearly varying bending moment which, at $x=L/4$ is equal to $-(3/4)PL$.

We note that the shear between $x=0$ and $x < L/4$ equals R_A, the reaction at the left end, and that the bending moment must return to zero. The discontinuity in the shear force at *B* allows the discontinuity in slope of M_b at that point.

Our *linear, ordinary, second order* differential equation for the deflection of the neutral axis becomes, *for x < L/4*

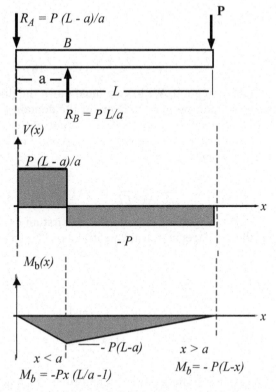

$$\frac{d^2v}{dx^2} = \frac{M_b(x)}{EI} = -\frac{3P}{EI} \cdot x$$

Integrating this is straight forward. I must be careful, though, not to neglect to introduce a constant of integration. I obtain

$$\frac{dv}{dx} = -\frac{3P}{EI} \cdot (x^2/2) + C_1$$

This is the slope of the deflected neutral axis as a function of *x*, at least within the domain $0<x<L/4$. Integrating once more produces an expression for the displacement of the neutral axis and, again, a constant of integration.

$$v(x) = -\frac{3P}{EI} \cdot (x^3/6) + C_1 \cdot x + C_2$$

Here then is an expression for the deflected shape of the beam in the domain left of the support at *B*. But what are the constants of integration? We determine the constants of integration by evaluating our expression for displacement $v(x)$ and/or our expression for the slope dv/dx at points where we are sure of their values. One such *boundary condition* is that, at $x=0$ the displacement is zero, i.e.,

$$v(x)\big|_{x=0} = 0$$

Another is that, at the support point *B*, the displacement must vanish, i.e.,

$$v(x)\big|_{x=L/4} = 0$$

These yield $C_2 = 0$ and $C_1 = \dfrac{PL^2}{(32EI)}$ and we can write, for $x < L/4$

$$v(x) = -\frac{P}{2EI} \cdot (x^3 - x/16)$$

So far so good. We have pinned down the displacement field for the region left of the support point at B. Now for the domain $L/4 < x < L$.

The *linear, ordinary, second order* differential equation for the deflection, again obtained from the moment/curvature relation for small deflections and rotations, becomes

$$\frac{d^2 v}{dx^2} = -\frac{P}{EI} \cdot (L - x)$$

Integrating this twice we obtain, first an expression for the slope, then another for the displacement of the neutral axis. To wit:

$$\frac{dv}{dx} = -\frac{P}{EI} \cdot (Lx - x^2/2) + D_1$$

and

$$v(x) = -\frac{P}{EI} \cdot [Lx^2/2 - x^3/6] + D_1 \cdot x + D_2$$

Now for some boundary conditions: It appears at first look that we have but one condition, namely, at the support point B, the displacement must vanish. Yet we have two constants of integration to evaluate!

The key to resolving our predicament is revealed by the form of the equation for the slope; we need to fix the slope at some point in order to evaluate D_1. We do this by insisting that the slope of the beam is continuous as we pass over the support point B. That is, the two slopes, that of $v(x)$ evaluated at the left of B must equal that of $v(x)$ evaluated just to the right of B. Our boundary conditions are then, for $x > L/4$:

$$v(x)|_{x = L/4} = 0 \qquad \text{and} \qquad \left(\frac{dv}{dx}\right)_{x = L/4} = -\frac{PL^2}{16EI}$$

where the right hand side of this last equation has been obtained by evaluating the slope to the left of B at that support point. Sparing you the details, which you are encouraged to plough through at your leisure, I - and I hope you - obtain

$$D_1 = (5/32) \cdot \frac{PL^2}{EI} \qquad \text{and} \qquad D_2 = -(1/96) \cdot \frac{PL^3}{EI}$$

So, for $L/4 < x < L$ we can write:

$$v(x) = -\left(\frac{PL^3}{96EI}\right) \cdot [1 - 15(x/L) + 48(x/L)^2 - 16(x/L)^3]$$

Setting x=L we obtain for the tip deflection:

$$v(L) = -(3/16) \cdot \frac{PL^3}{EI}$$

where the negative sign indicates that the tip deflects downward with the load directed downward as shown.

The process and the results obtained above prompt the following observations:

- The results are dimensionally correct. The factor $PL^3/(EI)$ has the dimensions of length, that is $FL^3/[(F/L^2)L^4] = L$.

- We can speak of an *equivalent stiffness* under the load and write

$$P = K\Delta \qquad \text{where} \qquad K = (16/3) \cdot \frac{EI}{L^3}$$

  E.g., an aluminum bar with a circular cross section of radius 1.0*in*, and length 3.0 *ft.* would have, with $I = \pi r^4/4 = 0.785$ *in*^{4}, and $E = 10 \times 10^6$ *psi*, *an equivalent stiffness of K=898 lb/inch*. If it were but one foot in length, this value would be increased by a factor of nine.

- This last speaks to the sensitivity of stiffness to length: We say "the stiffness goes as the inverse of the length cubed". But then, the stiffness is even more sensitive to the radius of the shaft: "it goes as the radius to the fourth power". Finally note that changing materials from aluminum to steel will increase the stiffness by a factor of three - the ratio of the E's.

- The process was lengthy. One has to carefully establish an appropriate set of boundary conditions and be meticulous in algebraic manipulations[2]. It's not the differential equation that makes finding the displacement function so tedious; it's, as you can see, the discontinuity in the loading, reflected in the necessity of writing out a different expression for the bending moment over different domains, and the matching of solutions at the boundaries of these regions that makes life difficult.

Fortunately, others have labored for a century or two cranking out solutions to this quite ordinary differential equation. There are reference books that provide full coverage of these and other useful formulae for beam deflections and many other things. One of the classical works in this regard is Roark and Young, FORMULAS FOR STRESS AND STRAIN, 5th Edition, McGraw-Hill, 1975. We summarize selected results as follows.

---

2. It took me three passes through the problem to get it right.

End-loaded Cantilever

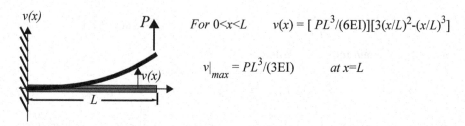

For $0 < x < L$ $v(x) = [PL^3/(6EI)][3(x/L)^2 - (x/L)^3]$

$v|_{max} = PL^3/(3EI)$ *at* $x = L$

Couple, End-loaded Cantilever

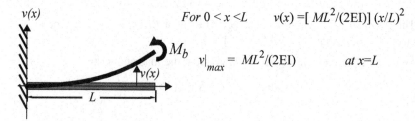

For $0 < x < L$ $v(x) = [ML^2/(2EI)](x/L)^2$

$v|_{max} = ML^2/(2EI)$ *at* $x = L$

Uniformly Loaded Cantilever

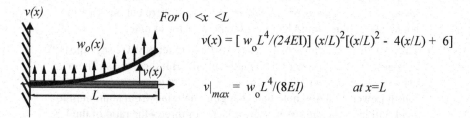

For $0 < x < L$

$v(x) = [w_o L^4/(24EI)](x/L)^2[(x/L)^2 - 4(x/L) + 6]$

$v|_{max} = w_o L^4/(8EI)$ *at* $x = L$

Uniformly Loaded Simply-Supported Beam

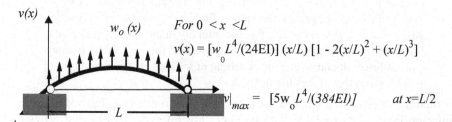

For $0 < x < L$

$v(x) = [w_o L^4/(24EI)](x/L)[1 - 2(x/L)^2 + (x/L)^3]$

$v|_{max} = [5w_o L^4/(384EI)]$ *at* $x = L/2$

Couple, End-loaded Simply-Supported Beam

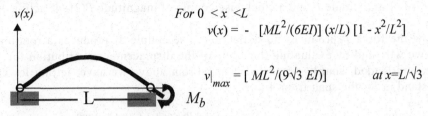

For $0 < x < L$

$$v(x) = - [ML^2/(6EI)] (x/L) [1 - x^2/L^2]$$

$$v|_{max} = [ML^2/(9\sqrt{3}\ EI)] \qquad at\ x = L/\sqrt{3}$$

Point Load, Simply-Supported Beam

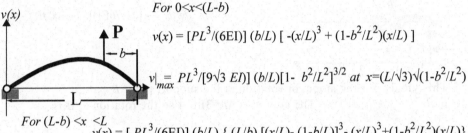

For $0 < x < (L-b)$

$$v(x) = [PL^3/(6EI)] (b/L) [-(x/L)^3 + (1-b^2/L^2)(x/L)]$$

$$v|_{max} = PL^3/[9\sqrt{3}\ EI)] (b/L)[1- b^2/L^2]^{3/2}\ at\ x = (L/\sqrt{3})\sqrt{(1-b^2/L^2)}$$

For $(L-b) < x < L$

$$v(x) = [PL^3/(6EI)] (b/L) \{ (L/b) [(x/L)- (1-b/L)]^3 - (x/L)^3 +(1-b^2/L^2)(x/L)\}$$

With these few relationships we can construct the deflected shapes of beams subjected to more complex loadings and different boundary conditions. We do this by *superimposing* the solutions to more simple loading cases, as represented, for example by the cases cited above.

Exercise 10.2

Show that the expression obtained for the tip deflection as a function of end load in the previous exercise can be obtained by superimposing the displacement fields of two of the cases presented above.

We will consider the beam deflection at the tip to be the sum of two parts: One part will be the deflection due to the beam acting as if it were cantilevered to a wall at the support point *B*, the middle figure below, and a second part due to the rotation of the beam at this imagined *root* of the cantilever at *B* —the figure left below.

We first determine the rotation of the beam at this point, at the support *B*. To do this we must imagine the effect of the load *P* applied at the tip upon the deflected shape back within the region $0 < x < L/4$. This effect can be represented as an equivalent force system at *B* acting internally to the beam. That is, we cut away the portion of the beam $x>L/4$ and show an equivalent vertical force acting downward of magnitude P and a clockwise couple of magnitude $P(3L/4)$ acting at B.

Now the force P produces no deflection. The couple M produces a rotation which we will find by evaluating the slope of the displacement distribution for a couple, end-loaded, simply supported beam. From above we have, letting lower case l stand in for the span from A to B,

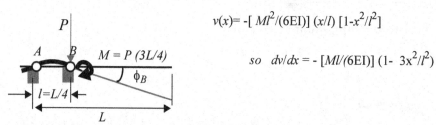

$$v(x)= -[Ml^2/(6EI)] (x/l) [1-x^2/l^2]$$

$$so \quad dv/dx = - [Ml/(6EI)] (1- 3x^2/l^2)$$

This yields, for the slope, or rotation at the support point B,
$dv/dx|_B = - Ml/(3EI)$. The couple is $M =3PL/4$ so the rotation at B is

$$\phi_B = -\frac{PL^2}{16EI}$$

The deflection at the tip of the beam where the load P is applied due to this rotation is, for small rotations, *assuming this portion rotates as a rigid body*, by

$$\Delta_{rigid\ body} = (3/4)\cdot L\phi_B = -\frac{3PL^3}{64EI}$$

where the negative sign indicates that the tip displacement due to this effect is downward.[3]

We now superimpose upon this displacement field, the displacement of a beam of length $3L/4$, imagined cantilevered at B, that is a displacement field whose slope is zero at B. We have, for the end loaded cantilever, that the tip displacement relative to the root is $-Pl^3/(3EI)$ where now the lower case "l" stands in for the length $3L/4$ and we have noted that the load acts downward. With this, the tip deflection due to this cantilever displacement field is

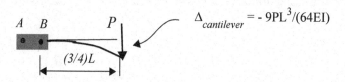

$$\Delta_{cantilever} = - 9PL^3/(64EI)$$

So the final result, the total deflection at the tip is, as before,
$$\Delta = \Delta_{rigid\ body} + \Delta_{cantilever} = - 3PL^3/(16EI).$$

---

3.   In this problem M is taken positive in opposition to our usual convention for bending moment. I have left off the subscript b to avoid confusion.

Exercise 10.3

Estimate the magnitude of the maximum bending moment due to the uniform loading of the cantilever beam which is also supported at its end away from the wall.

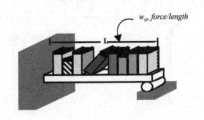

Deja vu! We posed this challenge back in an earlier chapter. There we made an estimate based upon the maximum bending moment within a uniformly loaded **simply supported** beam. We took $w_0 L^2/8$ as our estimate. We can do better now.

We use superpositioning. We will consider the tip deflection of a *uniformly loaded* cantilever. We then consider the tip deflection of an *end loaded* cantilever where the end load is just the reaction force (the unknown reaction force because the problem is statically indeterminate) at the end. Finally, we then sum the two and figure out what the unknown reaction force must be in order for the sum to be zero. This will be relatively quick and painless, to wit:

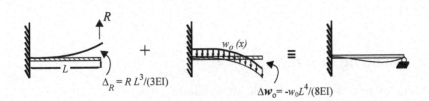

For the end-loaded cantilever we obtain $\Delta_R = R L^3/(3EI)$ where the deflection is positive up.

For the uniformly loaded cantilever we have $\Delta w_0 = -w_0 L^4/(8EI)$. The two sum to zero if and only if $\qquad R = (3/8) \cdot w_0 \cdot L$

We have resolved a statically indeterminate problem through the consideration of displacements and insisting on a deformation pattern compatible with the constraint at the end – that the displacement there be zero. With this, we can determine the reactions at the root and sketch the shear force and bending moment distribution. The results are shown on the next page.

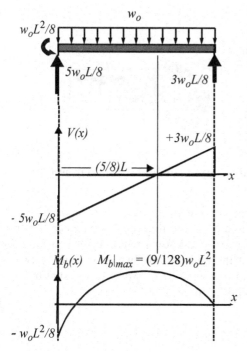

We see that there are two positions where the bending moment must be inspected to determine whether it attains a maximum. At the root we have $|M_b| = (1/8)w_oL^2$ while at $x=(5/8)L$ its magnitude is $(9/128)w_oL^2$. The moment thus has maximum magnitude at the root. It is there where the stresses due to bending will be maximum, there where failure is most likely to occur[4].

10.2 Buckling of Beams

Buckling of beams is an example of a failure mode in which relatively large deflections occur while no member or part of the structure may have experienced fracture or plastic flow. We speak then of *elastic buckling*.

Beams are not the only structural elements that may experience elastic buckling. Indeed *all* structures in theory might buckle if the loading and boundary conditions are of the right sort. The task in the analysis of the possibility of *elastic instability* is to try to determine what load levels will bring on buckling.

Not all elastic instability qualifies as failure. In some designs we **want** buckling to occur. The Tin Man's oil-can in The Wizard of Oz was designed, as were all oil cans of this form, so that the bottom, when pushed hard enough, dished in with a snap, *snapped through*, and displaced just the right amount of oil out the long conical nozzle. Other latch mechanisms rely upon *snap-through* to lock a fastener closed.

But generally, buckling means failure and that failure is often *catastrophic*. To see why, we first consider a simple mechanism of little worth in itself as a mechanical device but valuable to us as an aid to illustrating the fundamental phe-

---

4. If we compare this result with our previous estimate made back in exercise 3.9, we find the latter was the same magnitude! But this is by chance; In the simply supported beam the maximum moment occurs at mid-span. Here it occurs at the left end.

nomenon. The link shown below is to be taken as rigid. It is fastened to ground through a frictionless pin but a linear, torsional spring of stiffness K_T resists rotation, any deviation from the vertical. A weight P is suspended from the free end.

We seek all possible static equilibrium configurations of the system. On the left is the one we postulate working from the perspective of small deflections and rotations; that is **up until now we have always considered equilibrium with respect to the undeformed configuration**. We can call this a *trivial solution*; the bar does not rotate, the reaction moment of the torsional spring is zero, the rigid link is in compression.

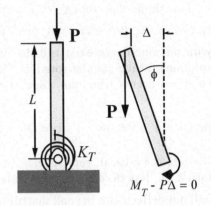

The one on the right is more interesting. But note; **we consider equilibrium with respect to a deformed configuration**.

Moment equilibrium requires that $M_T = P\Delta = PL\sin\phi$

or, assuming a linear, torsional spring ie., $M_T = K_T\phi$: $\left(\dfrac{K_T}{PL}\right)\cdot\phi - \sin\phi = 0$

Our challenge is to find values of ϕ which satisfy this equation for a given load P. We wish to plot how the load varies with displacement, the latter measured by either $\Delta = L\sin\phi$ *or* ϕ itself. That is the traditional problem we have posed to date. But note, this equation does not solve so easily; it is **nonlinear** in the displacement variable ϕ. Now when we enter the land of nonlinear algebraic equations, we enter a world where strange things can happen. We might, for example, encounter multi-valued solutions. That is the case here.

One branch of our solution is the trivial solution $\phi = 0$ for all values of P. This

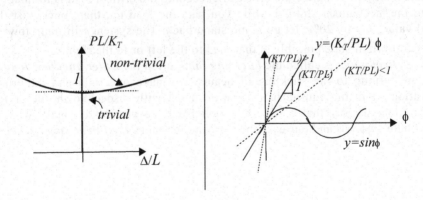

is the vertical axis of the plot below. (We plot the *nondimensional load* (PL/K_T) versus the *nondimensional horizontal displacement* (Δ/L). But there exists another branch, one revealed by the geometric construction at the right. Here we have plotted the straight line $y = (K_T/PL)\phi$, for various values of the load parameter, and the sine function $y = sin\phi$ on the same graph.

This shows that, for (K_T/PL) large, (or P small), there is no nontrivial solution. But when (K_T/PL) gets small, (or P grows large), intersections of the straight line with the sine curve exist and the nonlinear equilibrium equation has nontrivial solutions for the angular rotation. The transition value from no solutions to some solutions occurs when the slope of the straight line equals the slope of the sine function at $\phi = 0$, that is when $(K_T/PL)=1$, when

$$\boxed{\frac{PL}{K_T} = 1}$$

Here is a critical, very special, value if there ever was one. If the load P is less than K_T/L, **less than the critical value then we say the system is stable**. The link will not deflect. But beyond that all bets are off.

What is the locus of equilibrium states beyond the critical load? We can, for this simple problem, construct this branch readily. We find nontrivial solutions, pairs of (PL/K_T) *and* ϕ values that satisfy equilibrium most readily by choosing values for ϕ and using the equation to compute the required value for the load parameter. I have plotted the branch that results above on the left.

You can imagine what will happen to our structure as we approach and exceed the critical value; up until the critical load any deflections will be insensible, in fact zero. This assumes the system has no imperfections in the initial alignment of the link relative to vertical. If the latter existed, the link would show some small angular rotations even below the critical load. Once past the critical load we see very large deflections for relatively small increments in the load, and note the rigid bar could swing either to the right or to the left.

Another way to visualize the effect of exceeding the critical load is to imagine holding the mechanism straight while you take the load up, then, once past the critical value, say by 20%, let go... and stand back. The system will jump toward either equilibrium state possible at that load to the left or to the right[5].

The *critical value* is called the **buckling load**, often the **Euler buckling load**.

Before turning to the buckling of beams, we make one final and important observation: If in the equilibrium equation taken with respect to the deformed configuration, we say that ϕ, the deviation from the vertical is small, we can approximate $sin\phi \approx \phi$ and our equilibrium equation takes *the linear, homogeneous form*

---

5.   In reality the system would no doubt bounce around a good bit before returning to static equilibrium.

$$\left[1-\left(\frac{PL}{K_T}\right)\right]\cdot\phi = 0$$

Now how do the solutions of this compare to what we have previously obtained? At first glance, not very well. It appears that the only solution is the trivial one, $\phi = 0$. But look! I can also claim a solution if the bracket out front is zero. This will happen for a special value, an *eigenvalue*, namely when

$$PL/K_T = 1.0.$$

Now that is a significant result for, even though we cannot say very much about the angular displacement, other than it can be anything at all, we have determined the critical buckling load **without having to solve a nonlinear equation**.

We will apply this same procedure in our analysis of the buckling of beams. To do so we need to develop an equation of equilibrium for a beam subject to a compressive load that includes the possibility of small but finite transverse displacements. We do this by considering again a differential element of a beam but now allow it to deform before writing equilibrium. The figure below shows the deformed element acted upon by a compressive load as well as a shear force and bending moment.

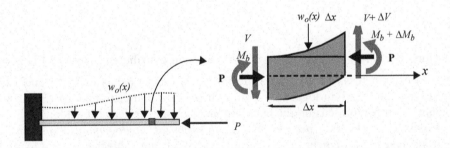

Force equilibrium in the horizontal direction is satisfied identically if we allow no variation of the axial load[6]. Force equilibrium of the differential element in the vertical direction requires

$$-V - w_o \cdot \Delta x + (V + \Delta V) = 0$$

while moment equilibrium about the station x yields

$$-M_b - w_o \cdot \frac{(\Delta x)^2}{2} + (V + \Delta V) \cdot \Delta x + (M_b + \Delta M_b) + P \cdot \Delta v = 0$$

---

6. The buckling of a vertical column under its own distributed weight would mean that P would vary along the axis of the beam.

The last term in moment equilibrium is only there because we have taken **equilibrium with respect to the** *slightly* **deformed configuration**. In the limit as $\Delta x \to 0$ we obtain the two differential equations:

$$\frac{dV}{dx} = w_o(x)$$

and

$$\frac{dM_b}{dx} + V + P \cdot \frac{dv}{dx} = 0$$

It is important to distinguish between the lower case and upper case *vee*; the former is the deflection of the neutral axis, the latter the shear force. We obtain a single equilibrium equation in terms of displacement by first differentiating the second equation with respect to x, then eliminate the term dV/dx using the first equation. I obtain

$$\frac{d^2 M_b}{dx^2} + P \cdot \frac{d^2 v}{dx^2} = -w_o(x)$$

We now phrase the bending moment in terms of displacement using the linearized form of the moment-curvature relation, $M_b/(EI) = d^2v/dx^2$ and obtain

$$(EI) \cdot \frac{d^4 v}{dx^4} + P \cdot \frac{d^2 v}{dx^2} = 0$$

Along the way I have assumed that the applied distributed load $w_0(x)$ is zero.

This is no great loss; it will have little influence on the behavior if the end load approaches the buckling load. It is a straight forward matter to take its effect into account. In order to focus on the buckling mechanism, we leave it aside.

This is a *fourth order, ordinary, linear differential, homogeneous equation* for the transverse displacement of the neutral axis of a beam subject to an end load, P. It must be supplemented by some boundary conditions on the displacement and its derivatives. Their expression depends upon the particular problem at hand. They can take the form of zero displacement or slope at a point, e.g.,

$$v = 0 \qquad \text{or} \qquad \frac{dv}{dx} = 0$$

They can also involve higher derivatives of $v(x)$ through conditions on the bending moment at a point along the beam, e.g., a condition on

$$M_b = (EI) \cdot \frac{d^2 v}{dx^2}$$

or on the shear force

$$V = -(EI) \cdot \frac{d^3 v}{dx^3} - P\frac{dv}{dx}$$

This last is a restatement of the differential equation for force equilibrium found above, which, since we increased the order of the system by differentiating the moment equilibrium equation, now appears as a boundary condition.

Exercise 10.4

A beam of Length L, moment of inertia in bending I, and made of a material with Young's modulus E, is pinned at its left end but tied down at its other end by a linear spring of stiffness K. The beam is subjected to a compressive end load P.

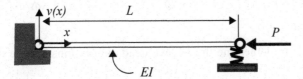

Show that the Euler buckling load(s) are determined from the equation

$$\left[\left(\frac{P}{KL}\right) - 1\right] \sin \sqrt{PL^2/(EI)} = 0$$

Show also that it is possible for the system to go unstable without any elastic deformation of the beam. That is, it deflects upward (or downward), rotating about the left end as a rigid bar. Construct a relationship for the stiffness of the linear spring relative to the stiffness of the beam when this will be the case, the most likely mode of instability.

We start with the general solution to the differential equation for the deflection of the neutral axis, that described by the function $v(x)$.

$$v(x) = c_1 + c_2 x + c_3 \sin \sqrt{\frac{P}{EI}} x + c_4 \cos \sqrt{\frac{P}{EI}} x$$

Letting $\lambda^2 = (P/EI)$ this can be written more simply as

$$v(x) = c_1 + c_2 x + c_3 \sin \lambda x + c_4 \cos \lambda x$$

In this, c_1, c_2, c_3, and c_4 are constants which will be determined from the boundary conditions. The latter are as follows:

At the left end, $x = 0$, the displacement vanishes so $v(0) = 0$ and since it is pinned, free to rotate there, the bending moment must also vanish $M_b(0) = 0$ or $d^2v/dx^2 = 0$ while at the right end, $x=L$, the end is free to rotate so the bending moment must be zero there as well: $M_b(L) = 0$ or $d^2v/dx^2 = 0$

The last condition (we need four since there are four constants of integration) requires drawing an isolation of the end. We see from force equilibrium of the tip of the beam that

$$V+F=0 \quad or \quad V+Kv(L) = 0$$

where F is the force in the spring, positive if the end moves upward, and V, the shear force, is consistent with our convention set out prior to deriving the differential equation for $v(x)$. Expressing V in terms of the displacement v(x) and its derivatives we can write our fourth boundary condition as:

$$d^3v/dx^3 + (P/EI)dv/dx - (K/EI) v(L) = 0 \quad or \quad d^3v/dx^3 + \lambda dv/dx - \beta v(L) = 0$$

where I have set $\beta = K/EI$.

To apply these to determine the c's, we need expressions for the derivatives of $v(x)$, up to third order. We find

$$dv/dx = c_2 + c_3\lambda \cos\lambda x - c_4 \lambda \sin\lambda x$$
$$d^2v/dx^2 = -c_3\lambda^2 \sin\lambda x - c_4 \lambda^2\cos\lambda x$$
$$d^3v/dx^3 = -c_3\lambda^3 \cos\lambda x + c_4 \lambda^3 \sin\lambda x$$

With these, the boundary conditions become

at $x= 0$.

$v(0)=0$: c_1 $+ c_4$ $= 0$

$d^2v/dx^2 = 0$: $+ c_4$ $= 0$

at $x = L$.

$d^2v/dx^2 = 0$: $(\lambda^2 \sin \lambda L) c_3$ $= 0$

$(\lambda^2 - \beta L)c_2$ $- (\sin \lambda L) c_3$ $= 0$

Now these are four, linear homogeneous equations for the four constants, c_1-c_4. One solution is that they all be zero. This, if your were to report to your boss would earn your very early retirement. The *trivial solution* is not the only one. In fact there are many more solutions but only for special values for the end load P, (λ). We know from our prior studies of systems of linear algebraic equations that the only hope we have for finding non zero c's is to have the determinant of the coefficients of the linear system be zero. The *eigenvalues* are obtained from this condition.

Rather than evaluate the determinant, we will proceed by an alternate path, no less decisive. From the second equation we must have $c_4 = 0$. Then, from the first we must have $c_1 = 0$.

Turning to the third equation, we might conclude that c_3 is zero as well. That would be a mistake. For we might have $\sin \lambda L = 0$.

Now consider the last two as two equations for c_2 and c_3. The determinant of the coefficients is

$$(\lambda^2 - \beta \cdot L) \cdot sin(\lambda L)$$

which, when set to zero, can be written

$$\left[\left(\frac{P}{KL}\right) - 1\right] sin\sqrt{PL^2/(EI)} = 0$$

This can be made zero in various ways.

- We can have

$$(P/KL) = 1$$

- or we can set

$$sin\,[PL^2/(EI)]^{1/2} = 0 \quad which\ has\ roots \quad PL^2/(EI) = \pi^2,\ 4\pi^2,\ 9\pi^2,...$$

The critical *eigenvalue* will be the lowest one, the one which gives the lowest value for the end load P. We see that this depends upon the stiffness of the linear spring relative to the stiffness of the beam as expressed by EI/L^3. For the mode of instability implied by the equation $(P/KL) = 1$, we must have

$$P = KL < \pi^2 EI/L^2 \qquad or \qquad K/(EI/L^3) < \pi^2$$

If this be the case, then the coefficient of c_2 in the last of our four equations will be zero. At the same time, $\sin \lambda L$ will *not* be zero in general so c_3 must be zero. The only non-zero coefficient is c_2 and, our general solution to the differential equation is simply

$$v(x) = c_2 x$$

This particular *buckling mode* is to be read as a *rigid body rotation* about the left end.

If, on the other hand, the inequality goes the other way, then another mode of instability will be encountered when

$$P = \pi^2 EI/L^2$$

Now c_2 must vanish but c_3 can be arbitrary. Our deflected shape is in accord with

$$v(x) = c_3\,sin\,\lambda x$$

and is sketched below. Note in this case the linear spring at the end neither extends nor contracts.

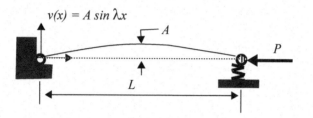

$$v(x) = A \sin \lambda x$$

Observe

- There are still other special, or eigenvalues, which accompany other, *higher*, mode shapes. The next one, corresponding to $PL^2/EI = 4\pi^2$, would appear as a full sine wave. But since the lower critical mode is the most probable, you would rarely see this what remains as but a mathematical possibility.

- If we let K be very large relative to (EI/L^3), we approach a beam pinned at both ends. The buckled beam would appear as in the figure above.

- Conversely, If we let K be very small relative to the beam's stiffness, we have a situation much like the system we previously studied, namely a rigid bar pinned at an end but restrained by a torsional spring. In fact, we get the same buckling load if we set the K_T of the torsional spring equal to KL^2.

- We only see the possibility of buckling if we consider **equilibrium with respect to the deformed configuration.**

10.3 Matrix Analysis of Frame Structures

We return to the use of the computer as an essential tool for predicting the behavior of structures and develop a method for the analysis of internal stresses, the deformations, displacements and rotations of structures made up of beams, beam elements rigidly fixed one to another in some pattern designed to, as is our habit, to support some externally applied loads. We call structures built up of beam elements, *frames*.

Frames support modern skyscrapers; your bicycle is a frame structure; a cantilevered balcony might be girded by a frame. If the structure's members are intended to support the externally applied loads via *bending,* the structure is a *frame*.

As we did with *truss* structures, structures that are designed to support the externally applied loads via *tension and compression* of its members, we use a *displacement method*. Our final system of equations to be solved by the machine will be the equilibrium equations expressed in terms of displacements.

The figure shows a frame, a building frame, subject to side loading, say due to wind. These structural members are not pinned at their joints. If they were, the frame would collapse; there would be no resistance to the shearing of one floor relative to another. The structural members are rigidly fixed to one another at their joints. So the joints can transmit a bending moment from one element to another.

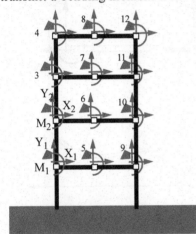

The figure at the left shows how we might model the frame as a collection of discrete, beam elements. The number of elements is quite arbitrary. Just how many elements is sufficient will depend upon several factors, e.g., the spatial variability of the externally applied loads, the homogeniety of the materials out of which the elements are made, the desired "accuracy" of the results.

In this model, we represent the structure as an assemblage of 16 elements, 8 horizontal, two for each floor, and 8 vertical. At each node there are three degrees of freedom: a horizontal displacement, a vertical displacement, and the rotation at the node. This is one more degree of freedom than appeared at each node of our (two-dimensional) truss structure. This is because the geometric boundary conditions at the ends, or junction, of a beam element include the slope as well as the displacement.

With 12 nodes and 3 degrees of freedom per node, our structure has a total of 36 degrees of freedom; there are 36 displacements and rotations to be determined. We indicate the applied external force components acting at but two of the nodes: X_1, Y_1 act at node 1 for example; M_1 stands for an applied couple at the same node, but we do not show the corresponding components of displacement and rotation.

Equilibrium in terms of displacement will require 36 linear, simultaneous equations to be solved. This presents no problem for our machinery.

We will construct the system of 36 equilibrium equations in terms of displacement by directly evaluating the entries of the whole structure's stiffness matrix. To do this, we need to construct the stiffness matrix for each individual beam element, then assemble the stiffness matrix for the entire structure by superimpositioning. What this means will become clear, I hope, in what follows.

Our approach differs from the way we treated truss structures. There we constructed the equilibrium equations by isolating each node of the truss; then wrote down a set of force-deformation relations for each truss member; then another matrix equation relating the member deformations to the displacement components of the nodes. We then eliminated the member forces in terms of these nodal displacements in the equilibrium equations to obtain the overall or global stiffness matrix for the entire truss structure. Only at this point, at the end of our construction, did we point out that each column of the stiffness matrix can be interpreted as the forces required to maintain equilibrium for a unit displacement corresponding to that column, all other displacements being held to zero. This is the way we will proceed from the start , now, constructing first the stiffness matrix for a horizontal beam element.

The figure at the right shows such an element. At each of its end nodes, we allow for an axial, a transverse force and a couple. These are assumed to be positive in the directions shown. The displacement components at each of the two ends - v_1, v_2, in the transverse direction, u_1, u_2, in the axial direction - and the slopes at the ends, ϕ_1 and ϕ_2, are also indicated; all of these are positive in the directions shown.

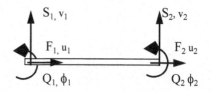

The bottom figure shows a possible deformed state where the displacements and rotations are shown more clearly (save u_2).

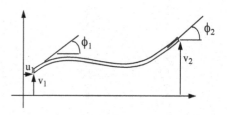

Our first task is to construct the entries in the stiffness matrix for this beam element. It will have the form:

$$\begin{bmatrix} F_1 \\ S_1 \\ Q_1 \\ F_2 \\ S_2 \\ Q_2 \end{bmatrix} = \begin{bmatrix} & & \\ & & \\ & & \\ & & \\ & & \\ & & \end{bmatrix} \begin{bmatrix} u_1 \\ v_1 \\ \phi_1 \\ u_2 \\ v_2 \\ \phi_2 \end{bmatrix}$$

A basic feature of matrix multiplication, of this expression, is the following: each column may be interpreted as the force and moment components, the left side of the equation, that are required to maintain a deformed configuration of a unit displacement corresponding to that column and zero displacements and rotations otherwise. For example, the entries in the first column may be interpreted as

F_1, S_1...Q_2 for the displacement
$$u_1 = 1 \quad \text{and} \quad v_1 = \phi_1 = u_2 = v_2 = \phi_2 = 0$$

This turns out to be a particularly easy column to fill in, for this deformation state, shown in the figure, only requires axial forces F1 and F2 to induce this displacement and maintain equilibrium.

F_1, u_1 F_2

In fact, if $u_1 = 1$, then the force required to produce this unit displacement is just (AE/L) 1. This means that F2, in this case, must, in order for equilibrium to hold, be equal to $- (AE/L)$. No other forces or moments are required or engendered. Hence all other elements of the first column of the stiffness matrix, the column corresponding to u_1, must be zero.

The elements of the fourth column, the column corresponding to the displacement u2, are found just as easily. In this case $F_2 = (AE/L)$ and $F_1 = - (AE/L)$ for equilibrium. Our stiffness matrix now has the form shown at the right. We continue considering the second column and envision the displace configuration

$$\begin{bmatrix} F_1 \\ S_1 \\ Q_1 \\ F_2 \\ S_2 \\ Q_2 \end{bmatrix} = \begin{bmatrix} \dfrac{AE}{L} & ? & ? & -\dfrac{AE}{L} & ? & ? \\ 0 & ? & ? & 0 & ? & ? \\ 0 & ? & ? & 0 & ? & ? \\ -\dfrac{AE}{L} & ? & ? & \dfrac{AE}{L} & ? & ? \\ 0 & ? & ? & 0 & ? & ? \\ 0 & ? & ? & 0 & ? & ? \end{bmatrix} \begin{bmatrix} u_1 \\ v_1 \\ \phi_1 \\ u_2 \\ v_2 \\ \phi_2 \end{bmatrix}$$

$v_1 = 1$ and $u_1 = \phi_1 = u_2 = v_2 = \phi_2 = 0$.

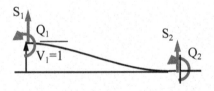

S_1

Q_1

$V_1 = 1$

S_2

Q_2

The figure at the left shows the deformed state. In this it appears we will require a S1 and a Q1 but no axial force. Remember, we assume small displacements and rotations so the axial force does not effect the bending of the beam element (and, similarly, bending of the beam does not induce any axial deformation). Only if we allow greater than small displacements and rotations will an axial load effect bending; this was the case in buckling but we are not allowing for buckling.

To determine what end force S_1, and what end couple, Q_1, are required to make the vertical displacement at the end of a cantilever equal 1 and the slope there zero - it's a cantilever because the vertical displacement and rotation at the right end must be zero -we make use of the known expressions for the tip displacement and rotation due to an end load and an end couple, then superimpose the two to ensure a vertical displacement of 1 and rotation of zero.

For a vertical, end load S_1, we have from the relationships given on page 182 ,

$$v_1 = \left(\frac{L^3}{3EI}\right) \cdot S_1 \qquad \text{and} \qquad \frac{dv}{dx}\bigg|_1 = \phi_1 = -\left(\frac{L^2}{2EI}\right) \cdot S_1$$

For an end couple, Q_1, we have

$$v_1 = -\left(\frac{L^2}{2EI}\right) \cdot Q_1 \qquad \text{and} \qquad \phi_1 = \left(\frac{L}{EI}\right) \cdot Q_1$$

If both a vertical force and a couple act, then, superimposing, we obtain the following two equations for determining the vertical displacement and the rotation at the end of the beam:

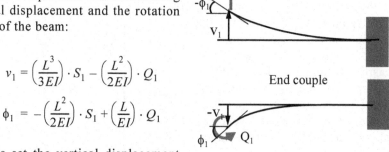

End load.

$$v_1 = \left(\frac{L^3}{3EI}\right) \cdot S_1 - \left(\frac{L^2}{2EI}\right) \cdot Q_1$$

$$\phi_1 = -\left(\frac{L^2}{2EI}\right) \cdot S_1 + \left(\frac{L}{EI}\right) \cdot Q_1$$

End couple

Now we set the vertical displacement equal to unity, $v_1 = 1$, and the slope zero, $\phi_1 = 0$, and solve these two equations for the required end load S_1 and the required end couple Q_1. We obtain:

$S_1 = (12EI/L^3)$ and $Q_1 = (6EI/L^2)$

The force and couple at the right end of the element are obtained from equilibrium. Without even drawing a free body diagram (living dangerously) we have:

$S_2 = -(12EI/L^3)$ and $Q_2 = (6EI/L^2)$

Thus the elements of the second column of the matrix are found and our stiffness matrix now has the form shown.

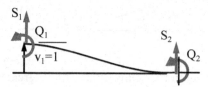

$$
\begin{bmatrix} F_1 \\ S_1 \\ Q_1 \\ F_2 \\ S_2 \\ Q_2 \end{bmatrix} =
\begin{bmatrix}
\dfrac{AE}{L} & 0 & ? & -\dfrac{AE}{L} & ? & ? \\[2mm]
0 & \dfrac{12EI}{L^3} & ? & 0 & ? & ? \\[2mm]
0 & \dfrac{6EI}{L^2} & ? & 0 & ? & ? \\[2mm]
-\dfrac{AE}{L} & 0 & ? & \dfrac{AE}{L} & ? & ? \\[2mm]
0 & -\dfrac{12EI}{L^3} & ? & 0 & ? & ? \\[2mm]
0 & \dfrac{6EI}{L^2} & ? & 0 & ? & ?
\end{bmatrix}
\begin{bmatrix} u_1 \\ v_1 \\ \phi_1 \\ u_2 \\ v_2 \\ \phi_2 \end{bmatrix}
$$

The elements of the third column, corresponding to a unit rotation $\phi_1 = 1$ and all other displacements and rotations zero, are found again from two simultaneous equations but now we find, with $v_1 = 0$ and $\phi_1 = 1$:

$S_1 = (6EI/L^2)$ and $Q_1 = (4EI/L)$

Again, the force and couple at the right end of the element are obtained from equilibrium. Now we have:

$$
\begin{bmatrix} F_1 \\ S_1 \\ Q_1 \\ F_2 \\ S_2 \\ Q_2 \end{bmatrix} =
\begin{bmatrix}
\dfrac{AE}{L} & 0 & 0 & -\dfrac{AE}{L} & ? & ? \\[2mm]
0 & \dfrac{12EI}{L^3} & \dfrac{6EI}{L^2} & 0 & ? & ? \\[2mm]
0 & \dfrac{6EI}{L^2} & \dfrac{4EI}{L} & 0 & ? & ? \\[2mm]
-\dfrac{AE}{L} & 0 & 0 & \dfrac{AE}{L} & ? & ? \\[2mm]
0 & -\dfrac{12EI}{L^3} & -\dfrac{6EI}{L^2} & 0 & ? & ? \\[2mm]
0 & \dfrac{6EI}{L^2} & \dfrac{2EI}{L} & 0 & ? & ?
\end{bmatrix}
\begin{bmatrix} u_1 \\ v_1 \\ \phi_1 \\ u_2 \\ v_2 \\ \phi_2 \end{bmatrix}
$$

$S_2 = -(6EI/L^2)$ and $Q_2 = (2EI/L)$

With this, our stiffness matrix becomes as shown.

Proceeding in a similar way at the right end of the beam element, we can construct the elements of the final two columns corresponding to a unit displacement, $v_2 = 1$ (the fifth colunm) and a unit rotation at the end, $\phi_2 = 1$, (the sixth column). Our final result is:

$$
\begin{bmatrix} F_1 \\ S_1 \\ Q_1 \\ F_2 \\ S_2 \\ Q_2 \end{bmatrix} = \begin{bmatrix} \dfrac{AE}{L} & 0 & 0 & -\dfrac{AE}{L} & 0 & 0 \\[2mm] 0 & \dfrac{12EI}{L^3} & \dfrac{6EI}{L^2} & 0 & -\dfrac{12EI}{L^3} & \dfrac{6EI}{L^2} \\[2mm] 0 & \dfrac{6EI}{L^2} & \dfrac{4EI}{L} & 0 & -\dfrac{6EI}{L^2} & \dfrac{2EI}{L} \\[2mm] -\dfrac{AE}{L} & 0 & 0 & \dfrac{AE}{L} & 0 & 0 \\[2mm] 0 & -\dfrac{12EI}{L^3} & -\dfrac{6EI}{L^2} & 0 & \dfrac{12EI}{L^3} & -\dfrac{6EI}{L^2} \\[2mm] 0 & \dfrac{6EI}{L^2} & \dfrac{2EI}{L} & 0 & -\dfrac{6EI}{L^2} & \dfrac{4EI}{L} \end{bmatrix} \begin{bmatrix} u_1 \\ v_1 \\ \phi_1 \\ u_2 \\ v_2 \\ \phi_2 \end{bmatrix}
$$

Exercise 10.5

Construct the stiffness matrix for the simple frame structure shown.

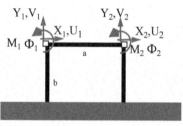

We employ the same approach as used in constructing the stiffness matrix for the single, horizontal beam element. We consider a unit displacement of each degree of freedom — U_1, V_1, Φ_1, U_2, V_2, and Φ_2 — constructing the corresponding column of the stiffness matrix of the whole structure in turn. Note how we use capital letters to specify the displacements and rotations of each node relative to a *global coordinate* reference frame.

We start, taking $U_1 = 1$, and consider what force and moment components are required to both produce this displacement and ensure equilibrium at the nodes. Since all other displacement and rotation components are zero, we draw the deformed configuration at the left. Refering to the previous figure, we see that we must have a horizontal force of magnitude $AE/a + 12EI/b^3$ applied at node #1, and a moment of magnitude $6EI/b^2$ to maintain this deformed configuration. There is no vertical force required at node #1.

At node #2, we see we must apply, for equilibrium of the horizontal beam element, an equal and opposite force to $X_1 = AE/a$. No other externally applied forces are required.

Thus, the first column of our stiffness matrix appears as shown at the right:

$$\begin{bmatrix} X_1 \\ Y_1 \\ M_1 \\ X_2 \\ Y_2 \\ M_2 \end{bmatrix} = \begin{bmatrix} \dfrac{AE}{a}+\dfrac{12EI}{b^3} & ? & ? & ? & ? & ? \\ 0 & ? & ? & ? & ? & ? \\ \dfrac{6EI}{b^2} & ? & ? & ? & ? & ? \\ -\dfrac{AE}{a} & ? & ? & ? & ? & ? \\ 0 & ? & ? & ? & ? & ? \\ 0 & ? & ? & ? & ? & ? \end{bmatrix} \begin{bmatrix} U_1 \\ V_1 \\ \Phi_1 \\ U_2 \\ V_2 \\ \Phi_2 \end{bmatrix}$$

Entries in the second column are obtained by setting V1 = 1 and all other displacement components zero. The deformed state looks as below;

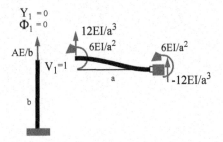

$Y_1 = 0$
$\Phi_1 = 0$
AE/b
$V_1 = 1$
$12EI/a^3$
$6EI/a^2$
a
b
$6EI/a^2$
$-12EI/a^3$

The forces and moments required to engender this state and maintain equilibrium are obtained from the *local* stiffness matrix for a single beam element on the previous page. Thus, the second column of our stiffness matrix can be filled in:

$$\begin{bmatrix} X_1 \\ Y_1 \\ M_1 \\ X_2 \\ Y_2 \\ M_2 \end{bmatrix} = \begin{bmatrix} \dfrac{AE}{a}+\dfrac{12EI}{b^3} & 0 & ? & ? & ? & ? \\ 0 & \dfrac{AE}{b}+\dfrac{12EI}{a^3} & ? & ? & ? & ? \\ \dfrac{6EI}{b^2} & \dfrac{6EI}{a^2} & ? & ? & ? & ? \\ -\dfrac{AE}{a} & 0 & ? & ? & ? & ? \\ 0 & -\dfrac{12EI}{a^3} & ? & ? & ? & ? \\ 0 & \dfrac{6EI}{a^2} & ? & ? & ? & ? \end{bmatrix} \begin{bmatrix} U_1 \\ V_1 \\ \Phi_1 \\ U_2 \\ V_2 \\ \Phi_2 \end{bmatrix}$$

Continuing in this way, next setting F1 = 1, all other displacements zero, sketching the deformed state, reading off the required force and moment components to maintain this deformed state and superimposing corresponding components at each of the two nodes, produces the stiffness matrix for the whole structure We obtain:

$$
\begin{bmatrix} X_1 \\ Y_1 \\ M_1 \\ X_2 \\ Y_2 \\ M_2 \end{bmatrix} =
\begin{bmatrix}
\dfrac{AE}{a}+\dfrac{12EI}{b^3} & 0 & \dfrac{6EI}{b^2} & -\dfrac{AE}{a} & 0 & 0 \\[2ex]
0 & \dfrac{AE}{b}+\dfrac{12EI}{a^3} & \dfrac{6EI}{a^2} & 0 & -\dfrac{12EI}{a^3} & \dfrac{6EI}{a^2} \\[2ex]
\dfrac{6EI}{b^2} & \dfrac{6EI}{a^2} & \dfrac{4EI}{b}+\dfrac{4EI}{a} & 0 & -\dfrac{6EI}{a^2} & \dfrac{2EI}{a} \\[2ex]
-\dfrac{AE}{a} & 0 & 0 & \dfrac{AE}{a}+\dfrac{12EI}{b^3} & 0 & \dfrac{6EI}{b^2} \\[2ex]
0 & -\dfrac{12EI}{a^3} & -\dfrac{6EI}{a^2} & 0 & \dfrac{AE}{b}+\dfrac{12EI}{a^3} & -\dfrac{6EI}{a^2} \\[2ex]
0 & \dfrac{6EI}{a^2} & \dfrac{2EI}{a} & \dfrac{6EI}{b^2} & -\dfrac{6EI}{a^2} & \dfrac{4EI}{a}+\dfrac{4EI}{b}
\end{bmatrix}
\begin{bmatrix} U_1 \\ V_1 \\ \Phi_1 \\ U_2 \\ V_2 \\ \Phi_2 \end{bmatrix}
$$

10.4 Energy Methods

Just as we did for Truss Stuctures, so the same perspective can be entertained for beams.

A Virtual Force Method for Beams

The intent here is to develop a way of computing the displacements of a (statically determinate) beam under arbitrary loading and just as arbitrary boundary conditions without making explicit reference to compatibility condiditions,   i.e. without having to integrate the differential equation for transverse displacements. The approach mimics that taken in the section on Force Method #1 as applied to statically determinate truss structures.

We start with the compatibility condition which relates the curvature, $\kappa = 1/\rho$, to the transverse displacement, $v(x)$,

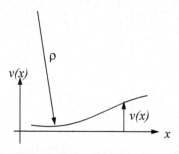

$$\kappa = \frac{d^2}{dx^2}v(x)$$

and take a totally unmotivated step, multiplying both side of this equation by a function of *x*, which can be anything whatsoever, we integrate over the length of the beam:

$$\int_0^L \kappa \cdot M^*(x) \cdot dx = \int_0^L M^*(x) \cdot \frac{d^2}{dx^2} v(x) \cdot dx$$

This arbitrary function bears an asterisk to distinguish it from the actual bending moment distribution in the structure.

At this point, the function *M** could be any function we wish, but now we manipulate this relationship, integrating the right hand side by parts and so obtain

$$\int_0^L \kappa \cdot M^*(x) \cdot dx = M^* \cdot \frac{dv}{dx}\Big|_0^L - \int_0^L \frac{dv}{dx} \cdot \frac{dM^*}{dx} \cdot dx$$

Integrating by parts once again, we have

$$\int_0^L \kappa \cdot M^*(x) \cdot dx = M^* \cdot \frac{dv}{dx}\Big|_0^L - \frac{d}{dx}(M^*) \cdot v\Big|_0^L + \int_0^L v \cdot \frac{d^2 M^*}{dx^2} \cdot dx$$

then consider the function M*(x) to be a *bending moment distribution*, **any** bending moment distribution that satisfies the equilibrium requirement for the beam, i.e.,

$$\frac{d^2}{dx^2} M^*(x) = p^*(x)$$

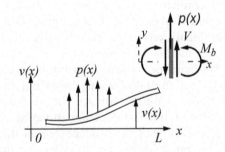

So p*(x) is arbitrary, because M*(x) is quite arbitrary - we can envision many different applied loads functions.

With this our compatibility relationship pre-multiplied by our arbitrary function, now read as a bending moment distribution, becomes

$$\int_0^L \kappa \cdot M^*(x) \cdot dx = M^* \cdot \frac{dv}{dx}\Big|_0^L - \frac{d}{dx}(M^*) \cdot v\Big|_0^L + \int_0^L p^*(x) \cdot v(x) \cdot dx$$

(Note: the dimensions of the quantity on the left hand side of this last equation are force times length or work. The dimensions of the integral and boundary terms on the right hand side must be the same).

Now we choose *p*(x)* in a special way; we take it to be a unit load at a single point along the beam, all other loading zero. For example, we take

$$p^*(x) = 1 \cdot \delta(x - a)$$

a unit load in the vertical direction at a distance *a* along the *x* axis.

Carrying out the integration in the equation above, we obtain just the displacement at the point of application, at *x* = *a*, i.e.,

$$v(a) = \int_0^L \kappa \cdot M^*(x;a) \cdot dx + M^* \cdot \frac{dv}{dx}\Big|_0^L - \frac{d}{dx}(M^*) \cdot v\Big|_0^L$$

We can put this last equation in terms of bending moments alone using the moment/curvature relationship, and obtain:

$$v(a) = \int_0^L \frac{M(x)}{EI} \cdot M^*(x;a) \cdot dx + M^* \cdot \frac{dv}{dx}\Big|_0^L - \frac{d}{dx}(M^*) \cdot v\Big|_0^L$$

And that is our special method for determining displacements of a statically determinate beam. It requires, first, solving equilbrium for the "actual" bending moment given the "actual" applied loads. We then solve *another* equilibrium problem - one in which we apply a unit load at the point where we seek to determine the displacement and in the direction of the sought after displacement. With this bending moment distribution determined from equilibrium, we carry out the integration in this last equation and there we have it.

Note: We can always choose the "starred" loading such that the "boundary terms" in this last equation all vanish. Some of the four terms will vanish because of vanishing of the displacement or the slope at a boundary. (The unstarred quantities must satisfy the boundary conditions on the actual problem). Granting this, we have more simply

$$v(a) = \int_0^L \frac{M(x)}{EI} \cdot M^*(x;a) \cdot dx$$

We emphasize the difference between the two moment distributions appearing in this equation; *M(x)*, in plain font, is the actual bending moment distribution in the beam, given the actual applied loads. *M*(x)*, with the asterisk, on the other hand, is some originally arbitrary, bending moment distribution which satisfies equilibrium - an equilibrium solution for the bending moment corresponding to a **unit load** in the vertical direction at the point *x=a*.

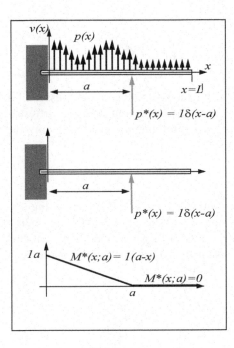

As an example, we consider a cantilever beam subject to a distributed load $p(x)$, which for now, we allow to be any function of distance along the span. We seek the vertical displacement at $x=a$.

We determine the bending moment distribution corresponding to the unit load at a. This is shown in the figure at the left, at the bottom of the frame. With this, our expression for the displacement at a becomes:

$$v(a) = \int_0^a \frac{M(x)}{EI} \cdot (a-x) \cdot dx$$

Note that:

i) M^* at $x=L$ and S^* $(= dM^*/dx)$ at $x=L$ are zero so the two boundary terms in the more general expression for $v(a)$ vanish.

ii) At the root, $v=0$ and f $(= dv/dx)$ $=0$ so the two boundary terms in the more general expression for $v(a)$ vanish.

iii) Finally, note that since the bending moment $M^*(x;a)$ is zero for $x>a$, the limit of integration in the above equation can be set to a.

We now simplify the example by taking our distributed load to be a constant, $p(x) = p_0$.

From the free body diagram at the right, we find

$$M(x) = p_0 \cdot (L-x)^2/2$$

and so the integral left to evaluate is:

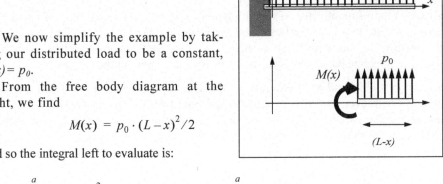

$$v(a) = \int_0^a \frac{p_0 \cdot (L-x)^2/2}{EI} \cdot (a-x) \cdot dx = \frac{p_0}{2EI} \cdot \int_0^a (L-x)^2 \cdot (a-x) \cdot dx$$

which, upon evaluation yields

$$v(a) = \frac{p_0}{24EI} \cdot (6a^2 L^2 - 4a^3 L + a^4)$$

Again, the significant thing to note is that we have produced an expression for the transverse displacement of the beam without confrontation with the differential equation for displacement! Our method is a force method requiring only the solution of (moment) equilibrium twice over.

A Virtual Displacement Method for Beams

The game now is to construct the stiffness matrix for a beam element using displacement and deformation/displacement considerations alone. We consider a beam element, uniform in cross-section and of length L, whose end displacements and rotations are presecribed and we are asked to determine the end forces and moments required to produce this system of displacements.

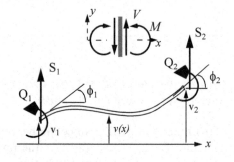

In the figure at the right, we show the beam element, deformed with prescribed end displacements v_1, v_2, and prescribed end rotations ϕ_1, ϕ_2. The task is to find the end forces, S_1, S_2 and end moments Q_1, Q_2, that will produce this deformed state and be in equilibrium and we want to do this without having to consider equilibrium explicitly.

We start with equilibrium:[7]

$$\frac{dV}{dx} = 0 \qquad \frac{dM}{dx} + V(x) = 0$$

and take a totally unmotivated step, multiplying the first of these equations, the one ensuring force equilbrium in the vertical direction, by some function $v^*(x)$ and the second, the one ensuring moment equilbrium at any point along the element, by another function $\phi^*(x)$, then integrate the sum of these products over the length of the element.

The functions $v^*(x)$ and $\phi^*(x)$ are quite arbitrary; at this stage in our game they could be anything whatsoever and still the above would hold true, as long as the shear force and bending moment vary in accord with the equilibrium requirements. They bear an asterisk to distinguish them from the actual displacement and rotation at any point along the beam element.

---

7. Note: our convention for positive shear force and bending moment is given in the figure.

$$\int_0^L \frac{dV}{dx} \cdot v^*(x) \cdot dx + \int_0^L \left[\frac{dM}{dx} + V(x)\right] \cdot \varphi^*(x) \cdot dx = 0$$

Now we manipulate this relationship, integrating by parts, noting that:

$$\int_0^L \frac{dV}{dx} \cdot v^*(x) \cdot dx = v^* \cdot V\big|_0^L - \int_0^L V(x)\frac{dv^*}{dx} \cdot dx$$

and

$$\int_0^L \frac{dM}{dx} \cdot \varphi^*(x) \cdot dx = \varphi^* \cdot M\big|_0^L - \int_0^L M(x)\frac{d\varphi^*}{dx} \cdot dx$$

and write

$$0 = v^* \cdot V\big|_0^L + \varphi^* \cdot M\big|_0^L - \int_0^L \left\{ V(x)\left[\frac{dv^*}{dx} - \varphi^*(x)\right] + M(x)\left[\frac{d\varphi^*}{dx}\right] \right\} \cdot dx$$

Now from the figure, we identify the internal shear force and bending moments acting at the ends with the applied end forces and moments, that is

$$S_1 = -V(0), \qquad Q_1 = -M(0), \qquad S_2 = +V(L), \qquad \text{and} \qquad Q_2 = +M(L).$$

so we can write

$$v_1^* \cdot S_1 + \varphi_1^* \cdot Q_1 + v_2^* \cdot S_2 + \varphi_2^* \cdot Q_2 = \int_0^L \left\{ V(x)\left[\frac{dv^*}{dx} - \varphi^*(x)\right] + M(x)\left[\frac{d\varphi^*}{dx}\right] \right\} \cdot dx$$

We now restrict our choice of the arbitrary functions $v^*(x)$ and $\phi^*(x)$[8]. We associate the first with a transverse displacement and the second with a rotation while requiring that there be no transverse shear deformation, i.e, plane cross sections remain plane and perpendicular to the neutral axis. In this case

---

8. Actually we have already done so, insisting that they are continuous to the extent that their first derivatives exist and are integrable, in order to carry through the integration by parts.

$$\frac{dv^*}{dx} - \varphi^*(x) = 0$$

so the first term in the integral on the right hand side vanishes, leaving us with the following;

$$v_1^* \cdot S_1 + \phi_1^* \cdot Q_1 + v_2^* \cdot S_2 + \phi_2^* \cdot Q_2 = \int_0^L M(x)\frac{d^2 v^*}{dx^2} \cdot dx$$

We can cast the integrand into terms of member deformations alone (and member stiffness, *EI*), by use of the moment curvature relationship

$$M(x) = EI \cdot \frac{d^2 v}{dx^2}$$

and write

$$v_1^* \cdot S_1 + \phi_1^* \cdot Q_1 + v_2^* \cdot S_2 + \phi_2^* \cdot Q_2 = \int_0^L EI \cdot \frac{d^2 v}{dx^2} \cdot \frac{d^2 v^*}{dx^2} \cdot dx$$

And that will serve as our special method for determining the external forces and moments, acting at the ends, given the prescribed displacement field *v(x)*. The latter must be in accord with equilibrium, our starting point; that is

$$\frac{d^2 M}{dx^2} = 0 \qquad \text{so from the moment/curvature relation} \qquad \frac{d^4 v}{dx^4} = 0$$

Thus *v(x)*, our prescribed displacement along the beam element, has the form:

$$v(x) = a_0 + a_1 \cdot x + a_2 \cdot x^2 + a_3 \cdot x^3 \quad \text{and} \quad \frac{d^2 v}{dx^2} = 2a_2 + 6a_3 \cdot x$$

With this, our relationship among end forces, end moments, and prescribed displacements becomes

$$v_1^* \cdot S_1 + \phi_1^* \cdot Q_1 + v_2^* \cdot S_2 + \phi_2^* \cdot Q_2 = \int_0^L EI \cdot (2a_2 + 6a_3 \cdot x) \cdot \frac{d^2 v^*}{dx^2} \cdot dx$$

The coeficients a_2 and a_3, as well as a_0 and a_1, can be related to the end displacements and rotations, v_1, ϕ_1, v_2 and ϕ_2 but we defer that task for the moment.

Instead, we choose a $v^*(x)$ to give a unit displacement $v_1^* = 1$ and the other end displacement and rotations to be 0. There are many functions that will do the job; we take

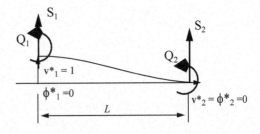

$$v^*(x) = \frac{1}{2} \cdot \left(1 + \cos\frac{\pi x}{L}\right)$$

differentiate twice, and carry out the integration to obtain $\quad S_1 = EI \cdot 6a_3$

Now we relate the a's to the end displacements and rotations. With

$$v_1 = v(0) = a_0$$

$$\phi_1 = \left.\frac{dv}{dx}\right|_0 = a_1$$

and

$$v_2 = v(L) = a_0 + a_1 \cdot L + a_2 \cdot L^2 + a_3 \cdot L^3$$

$$\phi_2 = \left.\frac{dv}{dx}\right|_L = a_1 + 2a_2 \cdot L + 3a_3 \cdot L^2$$

we solve for the a's and obtain

$$a_0 = v_1$$

$$a_1 = \phi_1$$

$$a_2 = -\left(\frac{3}{L^2}\right)v_1 - \left(\frac{2}{L}\right)\phi_1 + \left(\frac{3}{L^2}\right)v_2 - \left(\frac{1}{L}\right)\phi_2$$

$$a_3 = \left(\frac{2}{L^3}\right) \cdot v_1 + \left(\frac{1}{L^2}\right) \cdot \phi_1 - \left(\frac{2}{L^3}\right) \cdot v_2 + \left(\frac{1}{L^2}\right) \cdot \phi_2$$

which then yields the following result for the end force required, S_1, for prescribed end displacements and rotations v_1, ϕ_1, v_2, and ϕ_2.

$$S_1 = \left(\frac{12EI}{L^3}\right) \cdot v_1 + \left(\frac{6EI}{L^2}\right) \cdot \phi_1 - \left(\frac{12EI}{L^3}\right) \cdot v_2 + \left(\frac{6EI}{L^2}\right) \cdot \phi_2$$

The same ploy can be used to obtain the end moment Q_I required for prescribed end displacements and rotations v_I, ϕ_1, v_2, and ϕ_2. We need but choose our arbitrary function $v^*(x)$ to give a unit rotation $\phi_1{}^* = dv^*/dx = 1$ at the left end, $x=0$, and the other end displacements and rota-

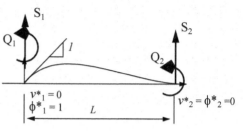

tions to be 0. There are many functions that will do the job; we take

$$v^*(x) = x - 2L(x/L)^3 + L(x/L)^5$$

differentiate twice, and carry out the integration to obtain $\qquad Q_1 = -EI \cdot 2a_2$

which then yields the following result for the end moment required, Q_I, for prescribed end displacements and rotations v_I, ϕ_1, v_2, and ϕ_2.

$$Q_1 = \left(\frac{6EI}{L^2}\right) \cdot v_1 + \left(\frac{4EI}{L}\right) \cdot \phi_1 - \left(\frac{6EI}{L^2}\right) \cdot v_2 + \left(\frac{2EI}{L^2}\right) \cdot \phi_2$$

In a similar way, expressions for the end force and moment at the right end, $x=L$ are obtained. Putting this all together produces the stiffness matrix:

$$\begin{bmatrix} S_1 \\ Q_1 \\ S_2 \\ Q_2 \end{bmatrix} = \begin{bmatrix} \dfrac{12EI}{L^3} & \dfrac{6EI}{L^2} & -\dfrac{12EI}{L^3} & \dfrac{6EI}{L^2} \\[2mm] \dfrac{6EI}{L^2} & \dfrac{4EI}{L} & -\dfrac{6EI}{L^2} & \dfrac{2EI}{L} \\[2mm] -\dfrac{12EI}{L^3} & -\dfrac{6EI}{L^2} & \dfrac{12EI}{L^3} & -\dfrac{6EI}{L^2} \\[2mm] \dfrac{6EI}{L^2} & \dfrac{2EI}{L} & -\dfrac{6EI}{L^2} & \dfrac{4EI}{L} \end{bmatrix} \begin{bmatrix} v_1 \\ \phi_1 \\ v_2 \\ \phi_2 \end{bmatrix}$$

Once again we obtain a symmetric matrix; why this should be is not made clear in taking the path we did. That this will always be so may be deduced from the boxed equation on a previous page, namely

$$\boxed{v_1^* \cdot S_1 + \phi_1^* \cdot Q_1 + v_2^* \cdot S_2 + \phi_2^* \cdot Q_2 = \int_0^L EI \cdot \frac{d^2v}{dx^2} \cdot \frac{d^2v^*}{dx^2} \cdot dx}$$

by choosing our arbitrary function $v^*(x)$ to be identical to the prescribed displacement field, $v(x)$.

We obtain:

$$v_1 \cdot S_1 + \phi_1 \cdot Q_1 + v_2 \cdot S_2 + \phi_2 \cdot Q_2 = \int_0^L EI \cdot (2a_2 + 6a_3x) \cdot (2a_2 + 6a_3x) \cdot dx$$

Now a_2 and a_3 may be expressed in terms of the prescribed end displacements and rotations via the relationships derived earlier, which, in matrix form, is

$$\begin{bmatrix} a_2 \\ a_3 \end{bmatrix} = \begin{bmatrix} -\dfrac{3}{L^2} & -\dfrac{2}{L} & \dfrac{3}{L^2} & \dfrac{2}{L} \\[2mm] \dfrac{2}{L^3} & \dfrac{1}{L^2} & -\dfrac{2}{L^3} & \dfrac{1}{L^2} \end{bmatrix} \begin{bmatrix} v_1 \\ \phi_1 \\ v_2 \\ \phi_2 \end{bmatrix} = [G] \begin{bmatrix} v_1 \\ \phi_1 \\ v_2 \\ \phi_2 \end{bmatrix}$$

Our basic equation then becomes:

$$v_1 \cdot S_1 + \phi_1 \cdot Q_1 + v_2 \cdot S_2 + \phi_2 \cdot Q_2 = EI \int_0^L \begin{bmatrix} v_1 & \phi_1 & v_2 & \phi_2 \end{bmatrix} [G]^T \cdot \begin{bmatrix} 4 & 12x \\ 12x & 36x^2 \end{bmatrix} \cdot [G] \begin{bmatrix} v_1 \\ \phi_1 \\ v_2 \\ \phi_2 \end{bmatrix} dx$$

so the symmetry is apparent.

If you carry through the matrix multiplication and the integration with respect to x you will recover the stiffness matrix for the beam element.

Design Exercise 10.1

Your task is to design a classroom demonstration which shows how the torsional stiffness of a structure depends upon material properties and the geometry of its stuctural elements.

Your professor has proposed the following as a way to illustrate torsional stiffness and, at the same time, the effects of combined loading on a shaft subjected to bending and torsion. A small number, N, circular rods or tubes are uniformly distributed around and fastened to two relatively rigid, end plates.

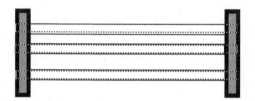

The rods are rigidly fixed to one of the end plates, say the plate at the left shown above. The other ends of the rods are designed so that they can be either free to rotate about their axis or not; that is, the right ends can be fixed rigidly to the plate or they can be left free to rotate about their axis.

If the rods are free to rotate about their axis, then all resistance to rotation of the entire structure is due to the resistance to bending of the N rods. (Note that not all N rods are shown in the figure). If, on the other hand, they are rigidly fixed at both ends to the plates, then the torsional stiffness of the overall structure is due to resistance to torsion of the N rods as well as bending.

A preliminary design of this apparatus is needed. In particular:

- The torsional stiffness of the entire structure due to bending of the rods is to be of the same order of magnitude as the torsional stiffness of the entire structure due to torsion of the rods.

- The overall torsional stiffness should be such that the rotation can be made visible to the naked eye for torques whose application does not require excess machinery.

- The apparatus should not fail, yield or break during demonstration.

- It should work with rods of two different materials.

- It should work with hollow tubes as well as solid shafts.

- Attention should be paid to how the ends of the rods are to be fastened to the plates.

Design Exercise 10.2

Back to the Diving Board

Reconsider the design of a diving board where now you are to rely on the elasticity of the board to provide the flexibility and dynamic response you desire. The spring at a will no longer be needed; a roller support will serve instead. In your design you want to consider the stresses due to bending, the static deflection at the end of the board, and its dynamic feel.

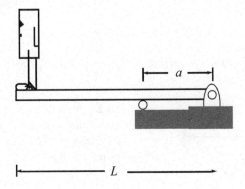

10.5 Problems - Stresses/Deflections, Beams in Bending

10.1   A force P is applied to the end of a cantilever beam but the end also is restrained by a moment M so that it can not rotate, i.e., the slope of the deflected curve is zero at *both* ends of the beam. The end *is* free to deflect vertically a distance Δ. We can write:

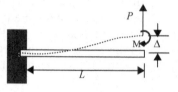

$P = K \Delta$

The beam is made from a material of Young's modulus E and its (symmetric) cross-section has bending moment of inertia I. Develop an expression for the stiffness K in terms of E, I and L, the length of the beam.

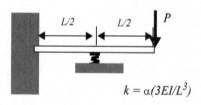

10.2   A cantilever beam is supported mid-span with a linear spring. The stiffness of the spring, k, is given in terms of the beam's stiffness as

$$k = \alpha(3EI/L^3)$$

• Determine the reactions at the wall, and the way the shear force and bending moment vary along the beam.

• Compare the tip deflection with that of a cantilever without mid-span support.

• What if α gets very large? How do things change?

• What if α gets very small? How do things change?

10.3   Determine the reactions at the three rollers of the redundantly supported beam which is uniformly loaded.

Sketch the shear force and bending moment distribution.

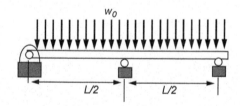

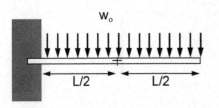

10.4 A cantilever beam carries a uniformly distributed load. Using Matlab, and the derived beam element stiffness matrix, (2 elements) determine the displacements at midspan and at the free end. Compare with the results of engineering beam analysis provided in the text.

In this, choose a steel beam to support a distributed load of 1000 lb/ft, and let the length be 20 ft.

Run Frameworks with 2, 3, 4 elements and compare your results.

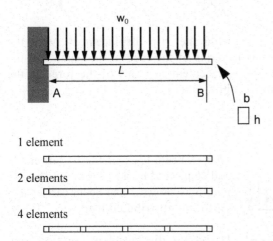

10.5 The cantilever beam AB carries a uniformly distributed load $w_0 = 31.25$ lb/in. Its length is $L = 40$ in. and its cross section has dimensions $b= 1.5$ in. & $h = 2$ in. Take the Elastic Modulus to be that of Aluminum, $E = 10$ E +06

a) Show that the tip deflection, according to engineering beam theory is

$v(L) = -1.0$ in

b) What is the beam deflection at mid-span?

c) Model the beam using Frameworks, in three ways; with 1, 2 and 4 elements. "Lump" the distributed load at the nodes in some rational way. Compare the tip and midspan deflections with that of engineering beam theory.

10.6 For the beam subject to "four point bending", determine the expression for the mid-span displacement as a function of P, L and a.

Do the same for the displacement of a point where the load is applied.

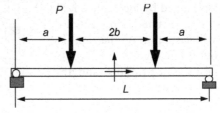

10.7   Given that the *tip deflection* of a cantilever beam, when loaded *mid-span*, is linearly dependent upon the load according to

$P_m = k_m \delta_m$ where k_m = 1000 N/mm

and given that the *tip deflection* of a cantilever beam, when loaded *at its free end*, is linearly dependent upon the load according to

$P_e = k_e \delta_e$ where k_e = 300 N/mm

and given that the deflection of a spring when loaded is linearly dependent upon the load according to

$F_s = k_s \delta_s$ where k_s = 500 N/mm

develop a compatibility condition expressing the   tip displacement (with the spring supporting the end of the beam) in terms of the load at mid-span and the force in the spring. Expressing the tip displacement in terms of the force in the spring using the third relationship above, show that

$$F_s = \frac{k_s}{k_m \cdot (1 + k_s / k_e)} \cdot P_m$$

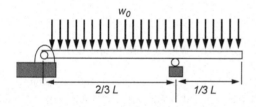

10.8 A beam is pinned at its left end and supported by a roller at 2/3 the length as shown. The beam carries a uniformly distributed load, w_0, <F/L>.

Derive the displacement function from the integration of the moment-curvature relationship, applying the appropriate boundary and matching conditions.

10.9   The roller support at the left end of the beam of the problem above is replaced by a cantilever support - i.e., its end is now fixed - and the roller support on the right is moved out to the end.

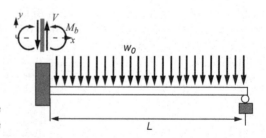

Using superpositioning and the displacement functions given in the text :

i) Determine the reaction at the roller.

ii) Sketch the bending moment distribution and determine where the maximum bending moment occurs and its value in terms of $w_0 L$.

iii) Where is the displacement a maximum? Express it in terms of $w_0 L^4 / EI$.

10.10 A beam, pinned at both ends, is supported by a wire inclined at 45 degrees as shown. Both members are two force members if we neglect the weight of the beam. So it is a truss. But because the beam is subject to an axial compressive load, it can buckle and we must analyze it as a beam-column.

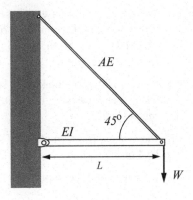

The wire resists the vertical motion of the end where the weight is applied just as a linear spring if attached at the end would do; so the buckling problem is like that of exercise 10.4 in the textbook.

If both the wire and the beam are made of the same material, (with yield stress σ_y):

i) Determine when the beam would buckle, in terms of the applied end load, W, and the properties of wire and beam. In this, consider all possible modes.

ii) What relationship among the wire and beam properties would have the wire yield at the same load, W, at which the beam buckles?

10.11 For small deflections and rotations but with equilibrium taken with respect to the deformed configuration, we derived the following differential equation for the transverse displacement of the end-loaded, "beam-column"

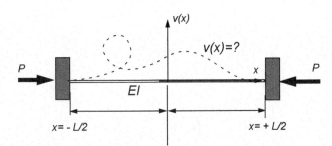

$$\frac{d^4 v}{dx^4} + \lambda^2 \cdot \frac{d^2 v}{dx^2} = 0$$

$$\text{where}$$

At x = $-L/2$ and x = $+L/2$, $v = 0$; $\frac{dv}{dx} = 0$

$$\lambda^2 = \frac{P}{EI}$$

The general solution is: $v(x) = c_1 + c_2 x + c_3 \sin\lambda x + c_4 \cos\lambda x$

Given the boundary conditions above, set up the eigenvalue problem for determining (1) the values of L for which you have a non-trivial solution (the eigenvalues) and (2) the relative magnitudes of the "c" coefficients which define the eigenfunctions.

10.12   We want to find an expression for the off-center displacement, *v(b)*, of the beam, Experiment #6, 1.105. We will do this in three ways:

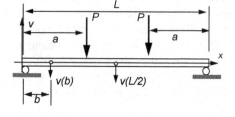

- By superpositioning the known solution for the beam carrying but a single (off-center) load.

- By the "virtual force" method of section 10.4 of the text.

- By a finite element computation using Frameworks.

We will take $L = 22$ in. $a = 8$ in. $b = 4$ in.   in what follows.

By superpositioning the known solution for the beam carrying but a single (off-center) load. For a point load, at distance a from the left end, we have

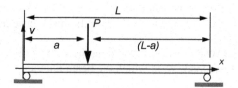

For x < a:

$$v(x) = \left(\frac{P(L-a)}{6LEI}\right) \cdot \left[-x^3 + \left\{L^2 - (L-a)^2\right\} \cdot x\right]$$

For x > a:

$$v(x) = \left(\frac{P(L-a)}{6LEI}\right) \cdot \left[\left(\frac{L}{(L-a)}\right) \cdot (x-a)^3 - x^3 + \left\{L^2 - (L-a)^2\right\} \cdot x\right]$$

Find the displacement at *x=b*, *v(b)*, for the case in which two loads are symmetrically applied. Express your results in the form $v(b) = (Some\ number) * PL^3/3EI$

10.13 By the "virtual force" method of section 10.4 of the text:

Here we have that

$$v(b) = \int\limits_{x=0}^{L} \frac{M_b(x)}{EI} \cdot \overset{*}{M}(x) \cdot dx$$

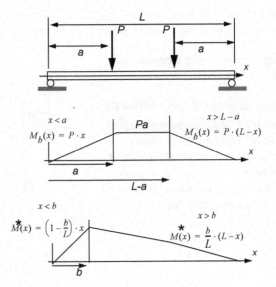

where $M^*(x)$ is the bending moment distribution due to a unit, virtual force acting at $x=b$. $M_b(x)$ is the real bending moment distribution due to the applied loads.

Because of the piecewise nature of the descriptions of these two functions of x, we must break the integration up into four parts:

We have, as shown in class (after correction of error):

$$v(b) = \frac{1}{EI} \cdot \left\{ \int_0^b \left(1 - \frac{b}{L}x\right) \cdot Pxdx + \int_b^a \frac{b}{L} \cdot (L-x) \cdot Pxdx \right\}$$

$$+ \frac{1}{EI} \cdot \left\{ \int_a^{L-a} \frac{b}{L}(L-x) \cdot Padx + \int_{L-a}^L \frac{b}{L} \cdot (L-x) \cdot P(L-x)dx \right\}$$

Also, as shown in class, we non-dimensionalize setting
$\xi = x/L$ $\alpha = a/L$ and $\beta = b/L$

This gives:

$$v(b) = \frac{PL^3}{EI} \cdot \left\{ (1-\beta)\int_0^\beta \xi^2 d\xi + \beta \left[\int_\beta^\alpha (1-\xi) \cdot \xi d\xi + \int_\alpha^{1-\alpha} (1-\xi) \cdot \alpha d\xi + \int_{1-\alpha}^1 (1-\xi)^2 d\xi \right] \right\}$$

Now alpha and beta are just numbers. So what is within the curly brackets is itself just a number; so once again we can express your results in the form

$v(b) = (Some\ number)*PL^3/3EI.$

Do this and compare with the first formulation.

10.14 By a finite element computation using Frameworks.

I suggest you use 6 elements as shown at the right.

If you choose *P, E* and *I* so that the factor PL^3/EI is some power (positive or negative) of ten, then the value of *v(b)* you obtain as output vertical displacement at node #1 will be easily compared with the two previous solutions.

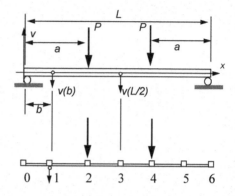

Physics

OPTICAL RESONANCE AND TWO-LEVEL ATOMS, L. Allen and J. H. Eberly. Clear, comprehensive introduction to basic principles behind all quantum optical resonance phenomena. 53 illustrations. Preface. Index. 256pp. 5⅜ x 8½.
0-486-65533-4

QUANTUM THEORY, David Bohm. This advanced undergraduate-level text presents the quantum theory in terms of qualitative and imaginative concepts, followed by specific applications worked out in mathematical detail. Preface. Index. 655pp. 5⅜ x 8½.
0-486-65969-0

ATOMIC PHYSICS (8th EDITION), Max Born. Nobel laureate's lucid treatment of kinetic theory of gases, elementary particles, nuclear atom, wave-corpuscles, atomic structure and spectral lines, much more. Over 40 appendices, bibliography. 495pp. 5⅜ x 8½.
0-486-65984-4

A SOPHISTICATE'S PRIMER OF RELATIVITY, P. W. Bridgman. Geared toward readers already acquainted with special relativity, this book transcends the view of theory as a working tool to answer natural questions: What is a frame of reference? What is a "law of nature"? What is the role of the "observer"? Extensive treatment, written in terms accessible to those without a scientific background. 1983 ed. xlviii+172pp. 5⅜ x 8½.
0-486-42549-5

AN INTRODUCTION TO HAMILTONIAN OPTICS, H. A. Buchdahl. Detailed account of the Hamiltonian treatment of aberration theory in geometrical optics. Many classes of optical systems defined in terms of the symmetries they possess. Problems with detailed solutions. 1970 edition. xv + 360pp. 5⅜ x 8½.0-486-67597-1

PRIMER OF QUANTUM MECHANICS, Marvin Chester. Introductory text examines the classical quantum bead on a track: its state and representations; operator eigenvalues; harmonic oscillator and bound bead in a symmetric force field; and bead in a spherical shell. Other topics include spin, matrices, and the structure of quantum mechanics; the simplest atom; indistinguishable particles; and stationary-state perturbation theory. 1992 ed. xiv+314pp. 6⅛ x 9¼.
0-486-42878-8

LECTURES ON QUANTUM MECHANICS, Paul A. M. Dirac. Four concise, brilliant lectures on mathematical methods in quantum mechanics from Nobel Prize-winning quantum pioneer build on idea of visualizing quantum theory through the use of classical mechanics. 96pp. 5⅜ x 8½.
0-486-41713-1

THIRTY YEARS THAT SHOOK PHYSICS: THE STORY OF QUANTUM THEORY, George Gamow. Lucid, accessible introduction to influential theory of energy and matter. Careful explanations of Dirac's anti-particles, Bohr's model of the atom, much more. 12 plates. Numerous drawings. 240pp. 5⅜ x 8½. 0-486-24895-X

ELECTRONIC STRUCTURE AND THE PROPERTIES OF SOLIDS: THE PHYSICS OF THE CHEMICAL BOND, Walter A. Harrison. Innovative text offers basic understanding of the electronic structure of covalent and ionic solids, simple metals, transition metals and their compounds. Problems. 1980 edition. 582pp. 6⅛ x 9¼.
0-486-66021-4

CATALOG OF DOVER BOOKS

A TREATISE ON ELECTRICITY AND MAGNETISM, James Clerk Maxwell. Important foundation work of modern physics. Brings to final form Maxwell's theory of electromagnetism and rigorously derives his general equations of field theory. 1,084pp. 5⅜ x 8½. Two-vol. set. Vol. I: 0-486-60636-8 Vol. II: 0-486-60637-6

MATHEMATICS FOR PHYSICISTS, Philippe Dennery and Andre Krzywicki. Superb text provides math needed to understand today's more advanced topics in physics and engineering. Theory of functions of a complex variable, linear vector spaces, much more. Problems. 1967 edition. 400pp. 6½ x 9¼. 0-486-69193-4

INTRODUCTION TO QUANTUM MECHANICS WITH APPLICATIONS TO CHEMISTRY, Linus Pauling & E. Bright Wilson, Jr. Classic undergraduate text by Nobel Prize winner applies quantum mechanics to chemical and physical problems. Numerous tables and figures enhance the text. Chapter bibliographies. Appendices. Index. 468pp. 5⅜ x 8½. 0-486-64871-0

METHODS OF THERMODYNAMICS, Howard Reiss. Outstanding text focuses on physical technique of thermodynamics, typical problem areas of understanding, and significance and use of thermodynamic potential. 1965 edition. 238pp. 5⅜ x 8½. 0-486-69445-3

THE ELECTROMAGNETIC FIELD, Albert Shadowitz. Comprehensive undergraduate text covers basics of electric and magnetic fields, builds up to electromagnetic theory. Also related topics, including relativity. Over 900 problems. 768pp. 5⅝ x 8¼. 0-486-65660-8

GREAT EXPERIMENTS IN PHYSICS: FIRSTHAND ACCOUNTS FROM GALILEO TO EINSTEIN, Morris H. Shamos (ed.). 25 crucial discoveries: Newton's laws of motion, Chadwick's study of the neutron, Hertz on electromagnetic waves, more. Original accounts clearly annotated. 370pp. 5⅜ x 8½. 0-486-25346-5

EINSTEIN'S LEGACY, Julian Schwinger. A Nobel Laureate relates fascinating story of Einstein and development of relativity theory in well-illustrated, nontechnical volume. Subjects include meaning of time, paradoxes of space travel, gravity and its effect on light, non-Euclidean geometry and curving of space-time, impact of radio astronomy and space-age discoveries, and more. 189 b/w illustrations. xiv+250pp. 8⅜ x 9¼. 0-486-41974-6

THE VARIATIONAL PRINCIPLES OF MECHANICS, Cornelius Lanczos. Philosophic, less formalistic approach to analytical mechanics offers model of clear, scholarly exposition at graduate level with coverage of basics, calculus of variations, principle of virtual work, equations of motion, more. 418pp. 5⅜ x 8½. 0-486-65067-7

Paperbound unless otherwise indicated. Available at your book dealer, online at www.doverpublications.com, or by writing to Dept. GI, Dover Publications, Inc., 31 East 2nd Street, Mineola, NY 11501. For current price information or for free catalogues (please indicate field of interest), write to Dover Publications or log on to www.doverpublications.com and see every Dover book in print. Dover publishes more than 400 books each year on science, elementary and advanced mathematics, biology, music, art, literary history, social sciences, and other areas.